ALASDAIR CAMERON
WELLHOUSE FARM
BLACK ISLE
MUIR OF ORD
ROSS-SHIRE IV6 7SF

The Countryman
Rescuing the Past

The Countryman
Rescuing the Past

Edited by ANN CRIPPS

DAVID & CHARLES : NEWTON ABBOT

0 7153 6071 X
© The Countryman 1973

Set in 11/13pt Plantin and printed in Great Britain
at The Pitman Press   Bath
for David & Charles (Holdings) Limited
South Devon House   Newton Abbot   Devon

Contents

Foreword

'Rescuing the Past' began as a regular feature in *The Countryman* in 1953. At that time I was Keeper of the recently founded Museum of English Rural Life at Reading University. John Cripps, who was then Editor, thought it would be a good way of stimulating the interest of readers in the many old tools and pieces of domestic and farm equipment that lie forgotten and disused in homes and on farms.

In the early issues we had to make use of a number of objects already presented to the Museum but the idea caught on very quickly and in no time readers were sending in drawings, descriptions and even objects (I was sometimes alarmed by the things that arrived in my post). In a way, almost more interesting and informative than the objects themselves were the many descriptions by people of how they had used them or seen them used, which often represented a real contribution to social history.

With some notable exceptions, historians have given little space to the run-of-the-mill work of house, farm and village and though it was sometimes quite easy to identify an object, it was quite another matter to discover exactly how it was used and the skills associated with it. In many cases, indeed, the same piece of equipment seems to have had a multitude of uses, some very far from the original intention.

Now, thanks to Ann Cripps's initiative, it is possible to see the fruits of 'Rescuing the Past' collected together in one volume—and what an excellent picture it presents of the ways of the country-man in the pre-industrial era. Many of the articles shown stretch back in history in almost unchanged form to the Middle Ages and even before, and represent the essential continuity of rural tradi-tions. This anthology of daily life and work reveals a wonderful picture of the industry, initiative and ingenuity—not to mention the drudgery and determination involved in earning a living—that characterised the country life of the past.

I don't think that any one of us—and 'Rescuing the Past' was very much a team effort both in the Museum and *The Countryman*—who was involved in compiling it ever expected that so fascinating a social 'documentary' of British rural life would emerge after twenty or so years.

When we began we knew that a great deal of material and a host of memories still remained available but time has taken its toll. The changes in the rural home and on the farm and in the village have been so embracing that little evidence of the pre-industrial, pre-plastic and pre-fabricated age now exists. In an age when so many of us live in surroundings that daily become more similar and more standardised, 'Rescuing the Past' may serve to inform and amuse us, but I hope that more than that it will provide a link in the continuous process of change and a link, too, in the traditions that will always be associated with earning a living from the land.

John Higgs

John Higgs is currently with the Food and Agriculture Organization of the United Nations.

Part One
LIFE ON THE LAND

Horse on the Road

Introduction

The countryman has bred horses of many kinds and for many purposes. When most English roads were mere tracks the merchant looked for a stocky, sure-footed horse that could carry a heavy bale along them. The invention of the spring at the end of the seventeenth century brought coaches and, as the roads improved, competition between the flying stages and mails became fast and furious. The demand was for a larger horse, long in the leg, that could gallop freely and eat up the miles. Nineteenth-century industrialisation produced the railways, outmoding coaches, and the horse was now needed to pull the heavy iron products of the age in port and railway yard. Thus the shire rose to prominence; and the agricultural labourer, squeezed off the land by one of the recurring depressions, was quick to find a town job if he had worked with horses.

Rumbler bells

When Robert Wilson moved into his farm at Bishopstone near Swindon at the beginning of the century he found rumbler bells for a team of horses in the harness room. They are decorated with engraved lines and bear the initials 'R. W.' for Robert Wells, the famous bell-founder. He started to make church, horse and sheep bells at Aldbourne in Wiltshire in 1760, and his mark was still being used in the nineteenth century. The iron ball rolling inside the sphere produced a rumble, a sound less percussive than that of the more musical clapper striking an open-mouthed bell. A team of horses bearing bells cast by Robert Wells, ringing two octaves, gave a pleasant warning to travellers that a great wagon, filling the lane, was round the next corner. The clapper bell was hung in an inverted box which, fringed and tasselled, lingered on without the bells, as decoration to be worn with brasses on special occasions. S. H. Wise was still using bells for purely practical purposes in west Oxfordshire in the 1930s. When ploughing, a horse wore a swinger on the poll-piece; but when the wagon went to the railway station to pick up lime or seed, the swingers were replaced by small bells, which produced a jingle rather than a Robert Wells crash. There was always the fear that the wagon might be benighted, and it carried no light. Laden packhorses took up most of a lane, so they too had to give notice of their presence. They wore small rumbler bells. The illustration (bottom right) also shows a swinger, mounted on the poll-piece. The bells were probably of some help in keeping the horses in a bunch and on the move.

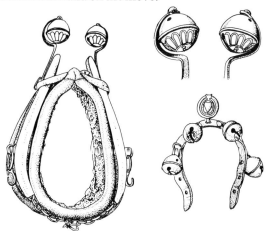

Casting a rumbler bell

Douglas Hughes of the Whitechapel Bell Foundry, which remains a craft shop where individual skill is cherished, told us that the foundry at Aldbourne in Wiltshire was one of four which the Whitechapel Foundry took over in the 1820s. 'Some of Wells's letter-stamps used on church bells, together with many of his patterns for small bells, were transferred at that time. Handbell ringing is popular today in the United States, and we have a busy export trade in bells cast, with one or two replacements, in Robert Wells's patterns. Recently we supplied a set of sleigh bells (eighteen straps each bearing rumbler bells tuned to a different note in the scale) for use in a fair-ground organ. The set of patterns covers eleven sizes, ranging in diameter from 1 to 3½in, and can be tuned to cover two chromatic octaves. There are several duplicates of pattern equipment to each size, so the demand for these bells or rumblers must have been heavy. The photographs (below and on page 16) show one bell casting and its relevant pattern equipment; and the sketches (right) illustrate, stage by stage, how a hollow bell containing a clapper is moulded. It is a remarkable example of advanced pattern making emanating from the eighteenth century.

'Moulding boxes, open at top and bottom, are arranged with pins so that they may be accurately located together. In the eighteenth century they would have been made of wood. A box is employed fully by using many patterns together, but for clarity I have shown only one. To ensure the precise positioning of the bell pattern an "oddside" cup (opposite) is used, so the first stage is to place the cup face down on a moulding board, put a box round it and fill this with moulding sand, which is rammed tightly; the upper surface is levelled off, as shown in Fig 1. The box is turned over on the board, and the surface of the sand well dusted with parting powder (bone flour) to ensure that further sand rammed on it will not stick. The rumbler-bell pattern, with sprue piece, is then placed in the oddside cup, and a fresh box is assembled and rammed up with sand as shown in Fig 2. The two boxes are turned over on the board, and the oddside (the first box with the cup rammed in it) is removed and stood to one side; it will be used again to start subsequent moulds. The lower box is then dusted with parting powder, and an empty box assembled on it, in place of the oddside, and rammed with sand to complete the outer mould (Fig 3).

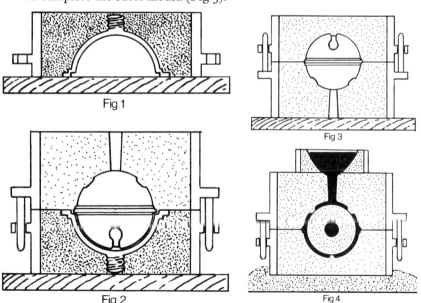

Fig 1

Fig 2

Fig 3

Fig 4

Training a young horse

The object below was found by D. Bird on his farm at Hepworth, near Diss. It is 9in long and made of ash. The peg which passes through both forks is attached to the main piece by a leather thong. It proved to be a tuttle box, used on the traces to keep two horses apart. The wooden pin which went through the two holes was either pointed or had the sharp end of a nail mounted on it, like the one from Norfolk. When the horses worked together one would be prodded by the spike. The box was secured to the trace chain. R. C. Lambeth, of Cambridge, wrote: 'These things are fortunately now illegal, and I really cannot understand any self-respecting horseman using them'.

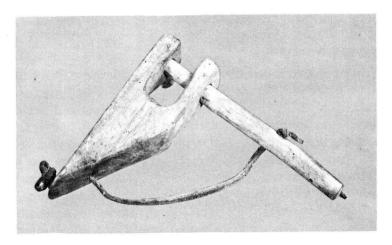

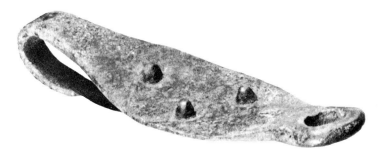

The bit attachment (opposite, lower) was used by a Suffolk horseman employed [1957] by the sender, John Vane, of Darmsden Hall, Needham Market. The three knobs were sharp-pointed and pricked the mouth of a young horse in training with an older one if it jerked away from its pair.

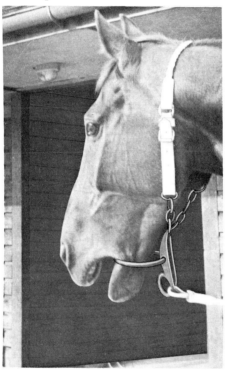

Stallion's bit

A bit identified, after a prolonged search, by two horse-breeders was used for leading stallions. The larger ring of the bit was put round the lower jaw, as in the accompanying photograph, the chain at the top of the bar being connected to a strap over the stallion's head, and a leading rein attached to the small ring at the bottom. If he tried to rear, pressure was put on the rein and the tip of the bar dug into the lower side of his jaw; the harder he reared the more it hurt. In Dentdale such a bit was called a 'chotter bit' or 'chofferin iron'.

Pack saddle

There were various ways of attaching a pack to a horse; for example,
pannier baskets might be thrown over them or loads be secured to
a pack saddle like the one above, which came from Totnes in
Devonshire. Made of wood with a straw-filled lining, it stands 25in
high.

Wagoner's belt

At the beginning of the century in Norfolk, belts were awarded to
young wagoners and ploughmen who won the approval of their

fellow-workers on an estate. As they were home-made, the pattern varied considerably. The belt shown below was given to N. Teulon Porter who wrote: 'In my early days, when I qualified as a wagoner on an estate in Norfolk I was awarded a terrific belt, which was made at home by one of my fellow-workers. Whether this was the last evidence of the custom in the county I cannot say, but I rather think that it was current on other large estates at the time'.

Stock on the Farm

Introduction

Sheep and oxen were companionable creatures, and readers recalled happy days spent with them. The shepherd who lived with a lambing flock, tending it day and night, had a happy task. From the reports which reached us he was contemplative and self-sufficient, only roused to anger when cows were given the roots he had marked for his flock.

Oxen pulling a plough were rarely shod, but those used to pull heavy loads from one part of the country to another left behind cues in fields into which they were turned during the night, and these can still be found today.

Work with oxen

George Ernest Holter, who lives at Seaford, was born at Exceat Farm in the cottage appearing distantly in the first of the old photographs, behind and to the right of the team of Sussex Reds. His eldest brother William, now eighty-seven, still lives there and works part-time on the farm. George was eleven years of age when he started work at Exceat Farm and, with a break for the 1914–18 war,

he stayed there until 1925, when Edward Percival Gorringe sold the oxen along with the farm. George's brother, Andrew, stands by the team led by Lamb and Leader, with Duke and Diamond following, and Quick and Nimble in the rear. The photograph of a Welsh Black team rolling was taken about 1905, and the boy at their head is another of the Holter brothers, Albert. George's wife Kathleen writes: 'There were probably two reasons why oxen were used quite extensively: they were easier to train than horses—the ox is by nature a quiet, gentle, plodding creature—and cheaper to feed, making do with hay, oat straw and a little cotton cake and dried grain.

'In Sussex the teams were yoked in pairs, and a chain ran from the middle of the yoke to the pole or neb of the wagon. When an old pair was retired, a young one had to be trained. A yoke would be held by a wall and the young oxen driven forward until they put their heads through the bows of curved ash that acted as collars. The youngsters were placed in the centre of the team, with an experienced pair before and behind. The master ox of a team was placed on the off side, and the oxherd called "Gee, Leader" for a left turn or "Hi, Lamb" for a right. If any persuasion was required, he used a 9ft ash stick or goad having at its tip a sharpened nail of the regulation length of one barleycorn; but a word was usually sufficient and, when ploughing, a trained team would turn at the end of a furrow without command.

'What was known as a pickaxe team was used for extra heavy work, such as pulling an old Sussex wagon full of large ($2\frac{1}{4}$cwt) sacks of grain from the thresher to the farm. Six oxen were harnessed in the lead to set a steady pace and share the strain with four horses behind. Even with such a team the oxen often sank up to their knees in the soft ground, their tails stiffening with the effort.

'The oxen did not need blinkers; but when they worked in tempting surroundings they wore nets over their muzzles, for if one had bent its neck in stopping to graze, it would have hurt the neck of its yoked partner. Ox cues were sometimes found on the farm, but even my father-in-law, who was born more than a hundred years ago, could not remember an ox shod in his time.'

In Gloucestershire, Oxfordshire and Wiltshire collars were preferred to yokes. 'Oxen broke themselves to harness', R. H. Wilson writes from Bishopstone Farm near Swindon. 'They were hitched to a baulk of timber and left in the yard. In three days they had learnt to pull.' J. R. Farmer of Filkins recalls: 'Wooden hames were used on horse and ox alike, but the ox hame has the "drawbar", the trace attachment, higher. The ox carried his neck flat and the line of the pull was along his backbone, whereas a horse pushes with shoulder and hock. A horse would put his weight against a load and, if he could not get it rolling, might give up; but an ox would move forward step by step and shift a heavier load than a couple of horses. I remember seeing the thresher coming to Oxlease: two horses pulled the engine, but the drum, which was heavier, was drawn by an ox, so slowly that you could hardly see it moving. The oxen were broken at eighteen months, worked two or three seasons and then, when the spring cultivations were done, fattened for Christmas. The meat was not so fat as some of the young stuff sold today. A great deal of their bulk was muscle put on at work'.

The famous team of Bathurst oxen out on the Gloucestershire road are wearing full 'spit and polish' harness with blinkers. The blinkers were worn as much to cut out distractions and temptations as to eliminate alarming sights.

R. H. Wilson talked of the old days with Dick Chivers, the last man to work oxen on his Bishopstone farm. 'Because of the horns the collar was U-shaped. It was put on upside down, fastened with a strap and then twisted round till the buckle and strap were on top and the thickest part of the flocking rested on the brisket or chest. We used wooden hames; our carpenter always had plenty of spares, and a break could be repaired or replaced in minutes. We had iron hames, but they were used chiefly for horses on road work, as a form of spit and polish: they usually had a little brass knob on top and a brass plate where the tug chain was attached. We worked two teams, each of four oxen, from 1904 until 1921.

'My father's chief excuse for this somewhat eccentric link with the past was that they kept the arable cultivation up to date. They were never called upon for hay-making or harvest; but when the horses were busy with these, the arable work would have fallen very much behind, had it not been for the continuous ploughing by the bullocks. The custom had one drawback; they became very clock-conscious, and nothing would persuade them to do another stroke of work after three o'clock, their recognised time for knocking off, unless the last turn would be pointing for home. On one occasion we needed one more wagon for hay, and my brothers and I thought that, if they were allowed to come back home and have their meal, we might be able to kid them to turn out again. To our surprise they offered no resistance, although it entailed two being fitted up with thill harness and put into the shafts of the wagon, a task they rarely encountered. All went well for a time, and we had nearly finished the load when they must suddenly have realised that they had been taken for a ride. At the far end of the field was a bridge spanning a considerable stream. Possibly urged on by the buzz of a gad-fly, off they went at full gallop, tails in the air, with myself and a brother on the top of the hay. The bullocks went under the bridge, but the load was too tall to follow them, so we jumped off into the stream. The bullocks remained in the shade, contentedly chewing the cud, until the cool of the evening, when we unhitched them and they walked quietly home, leaving the wagon and the load of hay.'

J. R. Farmer, at the age of ninety, was looking back at his long farming life: 'Tractors are lonely. I don't think I ever felt really well on one. There were the fumes, and the more there was to be done the faster you went and the worse it was. Working with oxen was cold slow work—trudge, trudge, trudge—and when they heard the 2.30 train go by, you couldn't get them to work no more. But horses, they were companionable. I had a pair and, except when I was marking out, I never had a rein on them. I just talked to them: "Come round" and "Gee there" '.

Drenching tools

The pewter drenching bottle (opposite) is 13in high and measures

almost the same round the base. It appears to have been brassed over
and has been soldered in several places. John L. Gilbert came across
it in 1940 at Wansford, Northamptonshire, where it had recently
been dredged from the River Nene. The Society of Pewter Collectors
has suggested that its date must be 1760–1800. The horn, from East
Hendred, Berkshire, was also used for dosing cattle and other
animals with liquid medicine. It is 8¾in long, and their teeth
marks can be seen in the reproduction. While a drench or 'ball' was
being administered to a horse, its mouth was kept open with a
farrier's gag. The one illustrated was made in a village smithy in
Kent and has a total length of 15½in. The pellet, which was shot
down the horse's throat with a balling gun, might measure as
much as 3in by 1in.

Bleeding instruments

At one time bleeding was a common form of veterinary treatment. Many people thought, for instance, that if a gallon of blood were taken from a stallion in spring it would quieten him down. The largest blade of the fleam shown (below right) was driven into a horse's jugular vein with a sharp blow from a fleam stick. The middle blade was used for cows and the smallest for calves. As soon as sufficient blood had been drawn the vein was closed with a pin. This fleam from Sevenhampton, Gloucestershire, is $3\frac{3}{4}$in long.

The scarifier or bleeding instrument (below left) from Miss D. Smith, of Charlton Kings. It belonged to her father Arthur William Smith, who was for forty-nine years a chemist at Pershore, Worcestershire, and was probably handed on to him by Francis Allen, for fifty years his predecessor in the same business. The scarifier was in common use in the late eighteenth and early nineteenth centuries for bleeding humans; indeed it was sometimes known as a mechanical leech, though this was in fact an instrument of another kind.

Dr Ashworth Underwood, of the Wellcome Historical Medical Museum, explained that the vertical lever was pulled over till it locked; this caused four blades, projecting below in the photograph, to disappear into slots. The bottom surface of the instrument was then applied to the skin and held firmly in position. The pressing of the button on the right released the blades, which flew across with great speed, making incisions in the skin. The depth of these was regulated by the screw on top of the instrument. Sometimes a cupping glass was then applied and the blood exhausted. The scarifier measures only 1½in across and is made of brass, with steel knives.

Apropos of this mechanical leech, Mrs Gwladys Tobit Huws sent interesting notes on the real thing. 'Fifty or sixty years ago', she wrote, 'leech doctors were in great demand among villagers who lived far from towns. In Cardiganshire one old woman used to supplement her meagre livelihood by collecting leeches. She would slip off her clogs, then her stockings, and wade into marshy ground. In a few minutes her legs would be covered with annelids, which fastened themselves on to her flesh and stuck as only a leech can do. She had to be very quick to free herself and transfer them to bottles. She supplied a retired doctor, who used them to relieve many common complaints, such as mumps or painful swellings (eg toothache), which defied other treatment. Just before the 1939–45 war a young naval man had a bad attack of earache. When it persisted he was sent to a Harley Street specialist, who gave him a prescription to take to a chemist. The latter told him to return the following day, as he had not the necessary ingredients. To the young man's surprise he was supplied with a bottle containing ten leeches, with instructions to put one in his ear daily, throwing it away as soon as it dropped. By the tenth day the ache had stopped entirely'. About 1907, when a relative of one of Mrs Huws's friends had ear trouble, a well-known Birmingham specialist prescribed similar treatment, and it was equally successful.

Teeth extractors

The outsize tooth extractor or key, 18in long, was found in a derelict veterinary establishment in Manchester. It is said to have been used for drawing the teeth of horses and cattle, in the same manner as those for extracting human teeth: when the block had been placed behind and the small swivel pushed against it, the tooth was removed by twisting and pulling. Below it is a blacksmith-made extractor used in rural dentistry to pull both human and horses' teeth.

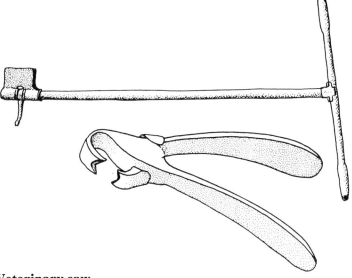

Veterinary saw

The photograph of a locally made saw with $4\frac{1}{2}$in bone handle came from John L. Gilbert. It belonged to a farrier at Stibbington near Peterborough during the last century. Professor Formston of the Royal Veterinary College told us that the instrument was used for making a transverse groove in the wall of a horse's hoof in the treatment of sand-crack.

Sheep bells

Shepherds claimed that bells kept a flock together; some said that they could tell by the sound whether the flock was disturbed, resting or grazing quietly. Others admitted that they had bells on their sheep solely for the pleasure of hearing them as they followed their lonely task of grazing the flock on solitary expanses of downland. The tonky-tonk of the canister was a pleasant sound, and canisters in graduated sizes, though not musically correct, were good enough to satisfy a shepherd who wanted a 'ring o' bells' on his flock. The clucket (below right) had a rounded crown and produced a chuckling sound – a 'clucket'. From New Zealand a reader, Robert Ahearn, inquired for a wether-bell. We passed the problem to T. H. G. Penny who knew shepherds on Salisbury Plain and 'the clonk-clonk of many-toned sheep bells far and near'. 'Bells were carried by the very best ewes only', he wrote. 'In a flock of three hundred only twenty-five or thirty might be belled, and their selection might take several hours. If a belled ewe deteriorated in any way, her bell would be removed and placed round the neck of another more worthy of the honour. A shepherd had only to glance at a belled ewe of another's flock to estimate its quality; and it was by her that a

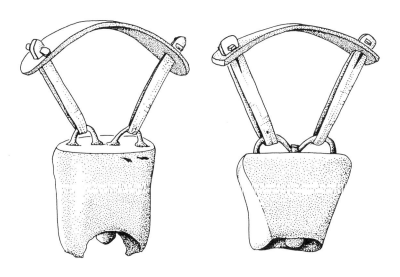

shepherd, too, was judged. The ewes were lambed in February and March. A certain number of chilvers (ewe lambs) were retained to strengthen the breeding flock. The rest of the chilvers and the wethers (castrated male lambs) were sent to the July and August sales, where the wethers were bought by butchers. A wether's life was seldom more than six months. If a shepherd had sent a particularly good pen of wethers to a sale, it was not uncommon for the auctioneer to compliment him with "You should have some wether-bells on those, shepherd". A really good wether was always known as a "bell-wether". A shepherd wishing to boast of his lambs might say that he could use a hundred wether-bells. I feel certain that "wether-bell" is only a figure of speech.' But from the fifteenth century a leader of sheep-like men was contemptuously called a bell-wether: 'the principal bell-wether of this mutiny'.

Field Work

Introduction

In the eighteenth and nineteenth centuries Britain was in ferment with the ideas of inventors, anxious to make the winning of food more efficient and less laborious, and manufacturers' catalogues are full of fascinating drills and ploughs—too many to list here. We have concentrated on the simple hand tools used to win food from the soil, tools that have their counterparts all over the world.

One of our most difficult tasks in running 'Rescuing the Past' was the selection of problem pictures for each instalment. If neither the Museum of English Rural Life nor we had any idea where or when an item had been used, it was impossible to say whether it was a 'one-off' made for a particular situation, and therefore unlikely to be identified, or an almost universal tool which would bring us letters from all over the English-speaking world.

Chappie

The chappie, an ancient implement, is still used by crofters in Wester Ross to break up the soil. They wield it like a hoe. Forged by a blacksmith, it is a survival of a primitive form of digging tool still found in many parts of the world. A similar tool with a much wider blade is used extensively in East Africa, where it is called a *jembe* and is swung well above the shoulder when digging. Some years ago in Uganda, John Higgs was assured that, though they were formerly made locally, most now come from Wolverhampton. Grubbing hoes or grubbers are commonly found in many parts of England but, unlike the chappie, they have flat ends. Estyn Evans describes a similar hoe, again with a flat end, used in Kerry and called a graffan.

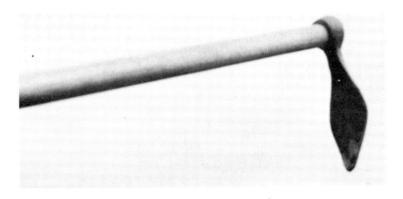

Foot plough

Standing opposite is a *cas-chrom* or foot plough found in the barn of a croft in Wester Ross. This implement was described in Sir John Sinclair's *Statistical Account of Scotland* (1791-9) thus: 'The *cas-chrom*, or crooked foot, is a crooked piece of wood, the lower end somewhat thick, about 2½ft in length, pretty straight, and armed at the end with iron, made thin and square to cut the earth. The shaft above the crook is generally straight, being 6ft long, and tapering upwards to the end, which is slender; just below the crook or angle, which is an obtuse one, there must be a hole, wherein a strong peg must be fixed for the workman's right foot, in order to push the instrument into the earth, while, in the meantime, standing upon his left foot, and holding the shaft firm with both hands; when he has in this manner driven the head far enough into the earth with one bend of his body, he raises the clod by the iron-headed part of his instrument, making use of the heel or hind part of the head as a fulcrum—in so doing turns it over always towards the left hand, and then proceeds to push for another clod in the same form'. The specimen illustrated lacks the sharp iron head with which it was once shod. Miss Speakman saw a young crofter using this implement on the shore of Loch Torridon, and he assured her that nothing else would be suitable for his rocky land. In a day a man can turn over about twice as much ground with a *cas-chrom* as with a spade.

Trace-link

Readers are often puzzled by 3in 'handles' found in the fields. They are spare trace-links which ploughmen and wagoners carried so that they could make speedy repairs to broken harness chains. A rather rusty example, in the state one would usually find a link, is shown below right.

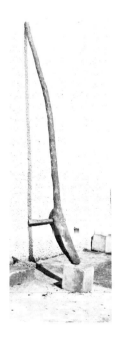
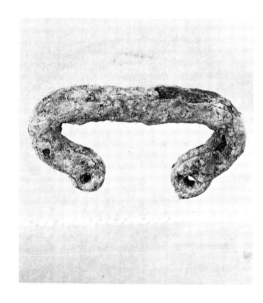

Forks and drags

Below left is a marling fork: a reminder of a once common practice on many British farms. Marl is a kind of soil consisting principally of clay mixed with carbonate of lime; it was added to light soils and was valuable as a fertiliser. In his *Epitome of the Art of Husbandry* published in 1675, writing on 'The Nature, Use and Benefit of Marle', Blagrave says, 'Some Countries yield Marle of several colours, as tis affirmed in Kent, wherein is found both yellow and gray; the blew and red are counted best'. The fork has widely spaced tines to prevent the heavy substance clogging. As shown here without its wooden handle it stands 30in high.

Above right is a hand-forged fork that was dug up in an unculti-vated orchard near Bromyard, Herefordshire, by Captain L. R. Hibbert. It measures 8¾in across, the centre tine is 6¾in long and the tool weighs 6lb. The handle has been provided to demonstrate the acute angle. This fork may have been used in hop growing. We asked if any reader had come across a similar tool whose original purpose was known to him. Several suggested that this was a drag for pulling farmyard manure from the back of a dung-cart to form heaps in the field, but the tines are only 6¾in long, whereas manure drags are quite heavy tools; the tines are usually at least 11in long. Nancy Peck, of Shanklin, told us that a similar tool, called a 'grapple', is used for breaking up clods in gardens in the Isle of Wight. As this is now difficult to buy new, people tend to take old forks to the blacksmith for bending. She adds that it is an admirable tool for preparing a seed bed on heavy soil; in Warwickshire, John Higgs had seen similar tools, called 'scratters', so used. John F. Pyott, of Stone, knows it as a Canterbury hoe, used in Staffordshire for hoeing potatoes. George Swinford, of Filkins, has used one for earthing up potatoes. To R. T. R. Barrett, of the Department of Agriculture, Malaya, it is an Assam fork, much used for digging up the roots of lalang, a persistent weed with rhizomes like couch-grass. M. Dengate of Sedlescome confirmed John Higgs's thought that the tool had been used in connection with hops. Her father, aged eighty-six, used to cultivate hops with a similar tool which he called a 'spiker' or spike hoe. E. P. Barker, of Tibberton, whose father and grandfather grew hops in Gloucestershire, writes that her father used such a fork for 'throwing down hops in the spring'; it seems that the hops were ploughed in the autumn and covered with soil, which was removed with this tool in spring. W. J. Loveless, of Otford, knows the tool as a Cambridge hoe, under which name it is sold in Kent today. His father used to hoe potatoes with it.

From John Peterson in Shetland we received the photograph below of a *tarigrep* or *taricrook*, now obsolete. The turned spike

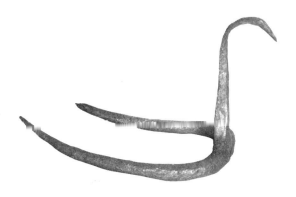

was driven into the wood near the end of a longish handle and
fastened, as in the sketch above, possibly with a strong lashing or a
metal ring. This tool was used to drag ashore floating seaweed for
manure. The word *tari*, of Norse origin, meant seaweed; today it
occurs only in the name of the implement and in certain place-
names.

Devil's wheelbarrow

'The Devil's wheelbarrow' is the name given by C. Winstone of
Upper Shaw Farm near Swindon to the putt illustrated below. A
two-wheeled lip cart for carrying muck was not uncommon in Devon,
Somerset, Wiltshire and Dorset, and it seems likely that this three-
wheeler came from the Bridgwater area of Somerset 40 years ago.
The putt or butt was also used for carting peat, and occasionally for
carrying to the top of a field the soil which had been washed to the
bottom. Seven feet long, it was hauled by a horse; the traces ran to
eyes on either side of the front wheel.

Wrought-iron ploughshare

The ploughshare above came from E. R. Bomford of Salford Priors Worcestershire. It is of wrought iron and appears to have been used on one of the long wooden mouldboard ploughs once common in that part of the country. John Higgs remembers one in use at Dorsington, near Stratford-upon-Avon, as recently as 1937. It was pulled by five horses in single line on very heavy clay land. The wooden mouldboard remained popular on clay, which did not stick to the wood. An unusual feature of the share illustrated is the fin coulter attached to its side. The share itself is 18in long and the coulter 8in high.

Harrow and plough cleaners

E. H. Ker asked for further information on the iron implement drawn below. It was found about 18in beneath the brick floor of a cellar in an old Thaxted inn. There is the usual socket for a wooden handle and an eye in the 18in head. As an apprentice C. E. Richards used a tool with a similar nose for guiding barrels of chemicals.

But the Thaxted tool weighs a hefty 2¾lb and the inner edge is flattened and strengthened. A blacksmith in the area, after handling it, had no hesitation in saying that it was used for lifting and cleaning harrows. He had made such tools in his youth, with 3ft wooden handles and tips well sharpened for pulling trash and squitch clear of the tines. The horses pulled a tree with a variable number of diamond-shaped frames attached, and the carter walking behind kept them clean, turning them over one by one. A farmer suggests that the eye was for tying the tool to the tree when not in use; and F. C. Brown points out that the hole in a butcher's cleaver enables it to be hung up to preserve the cutting edge. We are told that in west Oxfordshire harrows were cleaned with a tool like a short but heavy fish gaff.

Shortly after reading our paragraph on harrow cleaners Peter Rosser saw a similar tool on the fire of Mr Bendall, the blacksmith at St Mary Bourne, Hants, who knows it as a paddle used for clearing a plough clogged with stubble. For him the demand came with the combine-harvester and the long straw trash it left behind. During the 1939–45 war, when time was of the essence and the Ministry of Agriculture sent ploughing teams to short-handed farmers, he supplied them in some quantity.

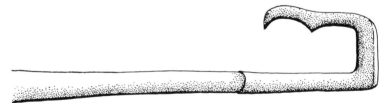

Broadcast sowing in England

Until the middle of the last century cereals and pulse were sown broadcast, wrote Dorothy Wright. In 1534, Master Fitzherbert gave an exact description of the method: 'Put thy peas into thy hopper and take a broad thong of leather or of garthewebbe [literally, girth webbing] of an ell long, and fasten it to both ends of the hopper and put it over thy head like a leash and stand in the midst of the land, where the sack lies, the which is most convenient for the filling of thy hopper, and set thy left foot before and take a handful of peas; and when thou takes up thy right foot then cast thy peas from thee all abroad; and when thy left foot riseth take another handful, and when thy right foot riseth cast them from thee. And so at every two paces thou shalt sow a handful of peas and so see that the foot and the hand agree and then shall ye sow even. And in your casting ye must open as well your fingers as your hand, and the higher and farther that you cast your corn the better shall it spread, except it be a great wind. And if the land be very good and will break small in the ploughing, it is better to sow after the plough than tarry any longer'.

Thorold Rogers, who examined many farm accounts, found no separate payment for sowing and thought it must have been done by the bailiff of a manor or by a farmer himself. But a document that describes farming at the time of Edward the Confessor names the sower among the other villagers and says that he ought to be paid with 'a seed-lip [hopper] full of every kind of every seed which he has properly sown throughout the space of the year'. Other villagers also ploughed and sowed for the lord of the manor.

Piers Ploughman both ploughed and sowed, and he gives some idea of the weight a sower carried: 'I will hang my seedbasket at the neck instead of a scrip and a bushel of breadcorn bring me therein'. The weight of a bushel measure of fine wheat seems always to have averaged about 60lb; barley and oats weigh considerably less. Some hoppers carried nearer two bushels. A full one was certainly very heavy, because a man or woman was said to be 'happered' or 'happerarsed': that is, shrunken or misshapen from the carrying of it. This is not to be wondered at, since sowing began in childhood.

Sowing is a common subject for illustration in medieval manuscripts because it is the symbol for October, and occasionally January, in English psalters and books of hours. The seed-lip or hopper was generally drawn in profile, and some were square, some rounded, some very large, some more like flower-baskets. In general, agricultural tools did not change much in shape until the coming of the machine, and not at all if they worked satisfactorily.

There was no competition then among makers to tempt people away from familiar tools.

The big basket of whatever shape was peculiarly English. Sometimes it was woven, sometimes coiled and sewn—a type of basket-making still called lip-work in some country areas. The sowing picture from Queen Mary's Psalter, which is fourteenth century, gives a choice of method; two sowers stand side by side, one with a seed-lip and one with a cloth.

The variety of seed-lip shapes was explained when, in the British Museum, I came on the Bohun Psalter dated about 1370. In two initials there are tiny figures of sowers, both carrying the same kind of basket. One, seen from the front, is kidney-shaped and the other, in profile, might be of any shape. This is not one of the great books, but the painter knew his subject. The expression of the taller man is melancholy, and he looks as if he was feeling the weight. So I believe that one reason for the apparent variety of this common farm tool in medieval times is that most artists suffered from the technical inability to draw a very difficult thing: a kidney shape, curved to the body, seen end on.

By the nineteenth century the seed-lip was better drawn (as in the 1802 illustration opposite) and perhaps better made, sometimes of wood, though I cannot think why, since a strong basket weighs far less and has a long life.

Fiddle

The fiddle saved neither time nor labour, but it did enable a less skilled man to cover the ground with seed evenly. The farmer photographed in Argyll (overleaf) trails a wooden marker from his belt as a further aid. Even in the seventies there are fiddles tucked away in barns 'just in case'. It comforts the farmer to think that in a really bad season he could sow the wet corner on foot with his fiddle, where a tractor would do nothing but harm.

Dibbles

Before the introduction of the drill, dibbles were used on farms, as they still are by gardeners, for planting all types of seed that were not broadcast. The specimen illustrated is a very simple tool shaped from a piece of thorn. It belonged to a farmer at Sudbury, Suffolk, is 12in long and was used for planting potatoes.

During the eighteenth and early nineteenth centuries there was, as E. G. Bolton pointed out, much argument among farmers in the Midlands and East Anglia over the merits of dibbling, which was then taking the place of broadcasting. Mr Ford of Long Melford wrote: 'Dibbling certainly has its advantages. The seed sown is reduced to about one bushel per acre: it takes a number of children from the spinning wheel to breathe the wholesome air in the fields, employed in dropping the grain in holes after the dibbler, and moreover, it appears such have equal if not superior crops to their neighbours who pursue the old plan'. A correspondent signing himself 'An Essex Farmer' affirmed that 'dibbling, as it respects the separation of the plants in the rows in which they stand, and the admission of air between them, is so far superior to drilling. The expense of extra labour is well borne in these times by the saving of seed'. On the other hand, a certain 'Farmer Sandy', who called himself a Northerner, had misgivings. In 1806 he wrote: 'Here we have to lament the frequency of adultery amongst our labourers in the field, but if we added dibble, dibbling, to our present practice, it would greatly increase. But in this part of the Kingdom, thanks to the discernment of our ancestors, our practice in aiding the poor is superior to that of England. I should therefore dread the introduction of dibble, dibble, dibbling, more from a regard to morality than from interested motives in respect to a great increase of poor's rates'.

Bird-scarer

The bird-scarer, from Winchcomb, Gloucestershire, measures 14in. Many old men alive today started work as crow-starvers or bird-minders, after the corn had been planted and again when it was in ear. The pay might be 2d a day, or 1d and a swede. Miss Hazel Inglis recalls one of the tongue-and-barrel type sixty years ago in Warwickshire. The farmer told her that the handle was made from the arm of his grandfather's chair, and the tongue from the busk of his grandmother's stays.

Weeding tools

The wooden tongs below left, 2ft 8in high and painted red, were found in the roof of an old house and sent by Mrs J. Wotherspoon from Hinton, near Rugby. They were used for pulling thistles, docks and other weeds from growing crops.

The 'ears' of the tool (opposite, lower right) are just $3\frac{1}{2}$in long and appear to have been sharpened on the insides, though the edges at the bottom of the 'V' are rounded. It was sent by Mrs J. S. Naish of Halesworthy, Suffolk who asked how it was used. William A. Cocks of Ryton-on-Tyne used a similar hand-made implement 'in the manner of a Dutch hoe to cut creeping thistle stems among growing corn. The inside edges were, as you say, sharpened and the implement went between the stalks quite easily'. A second use came from E. H. Ker: 'Fifty years ago an old man living in Gayton-le-Marsh, Lincolnshire, gave one to my father. It was described as an implement for throwing and roughly smoothing a mixture of mud, cut straw and cow dung on to a wooden lath structure when making and repairing mud-and-stud cottages. There were perhaps half a dozen of these dwellings in the village'. The sharpened inner edge would also have made it a suitable tool for chopping and mixing the daub. L. F. Salzman, the authority on medieval building, tells us: 'The probability is that workers on "mud-and-stud" improvised tools which would serve their purpose; these probably varied from village to village'.

Forks and hooks

The above photograph was taken at Filkins, a few miles from Burford, by W. T. Jones, of Preston Wynne, Hereford. The digger, which is about 15in high to the tread, belonged to Robert Prior, a farmer in the district, and had been used to dig the large white carrots which were at one time grown for stock-feeding. Relished by all classes of stock, the carrots gave a horse a fine coat and also enriched the colour of milk, cream and butter.

T. D. Parks sent the outline (1) of a tool found at Ash Green near Aldershot. There is an oast house on the estate, and it was possible that the 'fork', measuring 9½in in depth, was used to manoeuvre sacks as they swung on a rope hoist. H. T. Hooks told us it was a hop hook. The curled left-hand prong was used to cut the strings at the

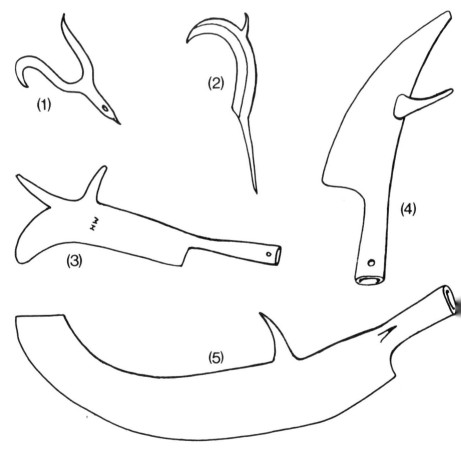

top of the pole and then the bine about 18in from the ground. With the crutch of the fork the plant was pushed up and over the pole, which was left standing in the ground. L. R. A. Groves, Curator of the Maidstone Museum, sent us sketch (2) of the conventional hop hook. It shows much less adaptation to the task and is close to (3), which was found in a copse seven miles from Dorchester. Claud Blair of the Victoria and Albert Museum, author of *European and American Arms*, writes of (3), with its curved cutting edge and two spikes: 'I have little doubt that this is not an agricultural implement but a fighting bill of the type used in enormous quantities in England during the 15th, 16th and 17th centuries'. J. E. Skyrme, who sent us the inquiry from Dorset, recalls that at the time of the Civil War Parliamentary troops were assembled at Hambledon Hill to mount an attack on the Royalists at Dorchester. There was little resistance or fighting, but Cromwell's army contained a large number of farmers and farm workers, many of whom deserted and made tracks for home. The bill was uncovered in a copse on the probable line of march from Hambledon Hill to Dorchester. It is possible that some of the country folk had their weapons made locally and rather roughly by smiths to the regulation pattern.

Tool (4) was found by R. Trevett at Nether Cerne. It was suggested that it was a peat knife; but J. E. Skyrme told us that this Dorset tool with convex blade and foot-rest, was a racing knife, 'for squaring up and clearing the "reens", "races" or water-channels. The course of the Cerne was flanked by water-meadows for most of its length; and the system of irrigation to induce prolific crops of grass for dairy cattle was widely used in this district up to the 1939–45 war. The tool would have been fitted with a short T-handle, like a spade'. The blade of (5) is 30in long and once had a hardened steel cutting edge. East Anglian thatching tools were offered at the same sale. We wondered if it might have been a knife for cutting hay from a rick, and the projection a foot rest. We had only one tentative suggestion for (5): that it was manure-knife rather than a hay-knife. The buries were cut into cubes and turned to hasten ripening; the weight (8lb) and sturdy construction would have been helpful in such a job.

When I went gleaning

No one in our Warwickshire village fifty years ago [in the 1890s] would have dreamed of leazing [gleaning] in a field which, though otherwise cleared of corn, still had a solitary stook somewhere in the middle of it. That stook told everybody who knew anything at all that the farmer had reason for wanting no leazers there yet awhile. Later on the stook would be carted away and anyone was free to glean.

Here again custom had to be observed. When a certain field was 'ready', able-bodied women, mothers and children took along refreshment and an assortment of old counterpanes, sheets and suchlike, to hold the grain we hoped to collect. There was no scramble for the most promising parts of the field. It was the rule that no one should begin until Mrs Fiddler, the village midwife and wise woman, gave the word. After exchanging greetings with the waiting company she would look round for stragglers. Yes, everbody was ready. Suddenly her arm would be flourished in a dramatic fashion, and she would cry, 'Now, then!'

We children each had a kind of satchel carried in front. 'If you won't work you shan't eat!' a mother would call shrilly to her inattentive offspring. 'Now, get busy! If you don't scrat you shan't peck—no, that you shan't!'

At that time two distinct methods of harvesting were common in our district. One was to cut the corn with hooks at ground level and bind into sheaves, as is done mechanically nowadays. The other way was to use a small sickle, with which a man would grasp perhaps half-a-dozen ears in his left hand and cut them off less than half-way down. This is why the sheaves shown in some old pictures are stumpy. They were easier to handle, store and thresh. Reaping this way was also less tiring. Leazing among the long straw stubble was not so pleasant as on the cleared fields. (About November, when autumn frosts had made the straw brittle, it would be knocked off with sticks and got into heaps. This was known as stubble-bagging.)

At last evening would come, and the big counterpane, now piled high with neat, small sheaves, would be knotted corner to corner, diagonally, and joisted on to a pram or mother's back. On the evening of the first leazing day furniture in our living room would be moved to make space in one corner for storing the grain. My father would lay the sheaves on the floor with butts in the corner and ears outward (so as to dry well and give mice the least chance), and pack them closely together in the form of an arc from wall to wall. The loose ears would be added at intervals as the stack rose higher. If the weather was fine and there were fields enough, our leazing would extend over several weeks.

After the farmer's stacks were threshed the leazer's corn would be put through the machine, free of charge. It was not uncommon for a cottager to have several bushels of leazed wheat. It was sent or taken to a picturesque, four-vaned windmill of the Dutch type. Again, no charge was made, but it was customary for the miller to keep the bran.—*S.H.S.*

In the Yard

Introduction

At harvest festival we still sing 'All is safely gathered in' and think of the old days when long hours were worked to snatch the crop before it was spoiled by the weather. A feeling of relief followed as the last waggon-load left the field to be stored in firmly thatched ricks and snug barns, and it is easy to forget how much still remained to be done.

Threshing was the most important and time-consuming job. There is still a specialised use for the flail and on a small scale the job is pleasant enough. But in earlier days it could mean long months of tedious labour. The steam-driven threshing drum was the great step forward. Feeding and tending this monster was hard work but it only stayed a couple of days before moving on to the next farm. Its arrival seems to have brought a welcome touch of drama to the quiet winter days. Today the whole process is completed inside the combine harvester as it moves steadily through the cornfield.

Stick and a half

Joe Firmin wrote: My father, who retired after forty years as a grower of vegetable and flower seeds in Essex, gave his flail to me with sadness in his face. He had just threshed his last batches of seeds with his 'old stick and a half', as he calls this sturdy hand implement.

One of my earliest memories of boyhood in north Essex is of watching my father and three workers flailing out seed in the old high-raftered and weather-boarded barn which adjoined our cottage. On a raw winter's day, when outside work was impossible, it was wonderful for a small boy to snuggle down in sweet-smelling straw or heaps of clean sacks, listening to the thump of the flails and watching the seed heads leap and fall at each round of blows. The flailers kept up a steady rhythm for hours on end. The effect was almost hypnotic on watcher and flailer, and the spell was broken only if one of the circle faltered and lost the beat, or the leader signalled a break.

My father remembers when most horticultural and agricultural stock seed crops in Essex and Suffolk were threshed by flail. As recently as the 1939–45 war, seed beans were knocked out by flails, as the mechanical threshers of those days tended to crack the skins of a

considerable percentage of the beans. Nowadays small crops of flowers and vegetables required for pedigree seed are still threshed by flail and dressed in hand blowers to avoid the risk of mixing valuable varieties and strains in large machines.

In the heyday of flailing, crops were taken into the barns when sufficiently dry. A floor was specially prepared for the operation, consisting of a three-inch covering of straw, over which was laid a cloth sheet with the crop to be threshed on top. The large doors at opposite ends of the barn were left open to provide a through draught for winnowing.

A flailing team could number up to six, but the most favoured number was four. The leader, known as the striker, started the threshing, followed by the others in rotation. This achieved the required rhythm, similar to the musical beat of 4/4 time. Some groups of flailers preferred to move slowly round the crop, gradually working towards the centre. Others formed a square and worked inwards.

If one was not pulling his weight, it was easily detected through the sound of the swingle. Older hands were so accurate that they gave exhibitions of their prowess and there were wagers on the result. My father and grandfather both told me of demonstrations which involved driving a four-inch nail into a piece of wood. The flailer would take a few ranging strokes with the swingle and then, with a sharp swing, knock the nail into the wood without bending it.

Each crop required a different weight of threshing. Before starting, the striker would indicate how much pressure should be used. After a few minutes the seed would be examined to make sure that the strokes were not too heavy.

The handle of a flail is usually made of ash or hazel and is 5ft or 5ft 6in long. The swingle, ideally, is half the length of the handstaff and made from ash, holly, blackthorn or apple. Old hands, my father included, used to like their swingles fairly knotty, and it might be ten years before these were replaced. The swingle is attached to the handle by a leather thong, which is joined to a shouldered loop (or cap) of willow or ash at the head of the handle.

The cap is made while the wood is green. I remember my

grandfather fashioning several at a time, chiselling the wood to the required boxed shape, then gently turning the tops in steam or boiling water. These caps, designed to act as swivels, are bound in the required position and dried for up to a year before use. By this time they are hard as bone. If care is not taken when the two open sides are brought together, the loop may crack.

By the 1850s most wheat was threshed mechanically. Barley continued to be flailed through to the close of the century because of lingering fears that machines injured malting quality. Sentiment, combined with a limited practical use, accounts for the survival of the 'stick and a half' among specialist seed growers in north and central Essex and in Suffolk.

Threshers

When unbruised straw was needed for 'Devon reed' thatching a
flail could not be used for the threshing, and small handfuls of ears
were knocked against the bars of the frame to shake out the grain.
The threshing frame (on the previous page), little more than 2ft in
height, comes from Devon. The early threshing machine (below),
photographed in a barn at St Clement, near Truro, has a rotating
drum with bars fixed to the outer surface, and a hinged lid.

Winnowers

The next task was to separate the chaff from the grain. This early
winnowing machine from Devizes, Wiltshire (opposite) was used
before threshers were fitted with winnowing devices. Frequently,
of course, the job was done on a windy day, but a simple contrivance
like this one from Devizes, was also useful. One man turned
the handle fast, while another threw the grain and chaff up into
the draught caused by the sacking. In any form of hand winnowing the lightest corn fell at the tail of the heap and thus came

to be know as 'tail corn'. The winnowing fan (below left) was
photographed by C. Henry Warren on a small farm at the back of
his house at Finchingfield in Essex. Few of these are now in existence, and they appear to have been used mainly in the eastern
counties of England for throwing up the grain and chaff in winnowing. The chaff was blown away while the corn fell back into the fan.
Miss Murray Speakman's grain sieve (below right) comes from
Scotland. Eighteen inches across, it is made of raw hide and wood
and was formerly used for winnowing.

Mills

The handmill dated 1608 was offered for sale at an auction in the Wiltshire village of Ham. The stones measure 18in across and were turned by two men. There can be few such mills of the period still in existence. This one seems to have been taken to the village carpenter for repair and lain there till the shop was sold.

Below, is a mill bill originally in use at Aythorpe Roding, Essex, and sent by F. W. Steer. It has a metal bit 9in long and was used to recut the feather pattern in the face of a millstone—a skilled operation which few men can now perform. The photograph (opposite left) shows a dresser at work in the 1950s on the runner stone of the Union windmill at Cranbrook, Kent. On the right the runner stone, resting on wedges, is about to be lowered on to the bedstone ready to grind corn again.

The travelling corn mill below was made at Romsey, Hampshire, about 1860 and operated by a contractor in the county. It was pulled and driven by a steam engine. The mill has two large stones similar to those used in a normal mill and also incorporates an oat crusher and a flour boulter.

The iron-tanged object (below) was photographed by G. E. Marston in Norfolk. It is a calder fork and was used to clear the cavings (broken straw and ears) from under a threshing-drum. 'Colder' or 'calder' is an East Anglian dialect term for cavings. Originally a long wooden handle was attached to the short central spike. Today most threshing-drums are fitted with boards which divert the cavings to an elevator, where they are mixed with the straw; but in the past they were kept apart for mixing with sliced roots for stock feed. The overall length of the tines is 2ft 9in, and I am assured that, although they are 12in apart, the fork was an effective tool.

Chaff-cutter and gorse-chopper

Cutting chaff with one of the old chaff-boxes was a complicated business. The straw was scraped together with a fork grasped in the left hand, held down by a clamp operated with the foot, and cut with the blade which was worked by the right hand. Men who specialised in chaff-cutting went round from farm to farm carrying on their backs boxes like that shown below left, which stands about 3ft high and came from Burwell Fen in Cambridgeshire.

Until the late nineteenth century, especially in the poorer parts of the country, gorse was fed to livestock; but first it had to be bruised and broken in order to make it edible. Several times during its early years the Royal Agricultural Society (founded 1838) offered prizes at its annual shows for machines for the job. John Higgs had not seen a hand tool for the purpose until Ivor E. Davies, of Penmaenmawr, Caernarvonshire, sent this head of a gorse-chopping mallet (bottom right opposite). His interest in the history of feeding gorse to stock led him to address an inquiry to a North Wales weekly paper, and it brought him the tool from H. T. Jones, of Talybont, who had seen it in use. The iron head is mounted on a round wooden mallet, half of which is now gone. The handle fitted into the hole of which part can be seen at the bottom right-hand corner. The mallet now stands 7in high.

Turnip-slicer and kibbler

The metal object in Mary Arden's house at Wilmcote puzzled Levi Fox, Director of the Shakespeare Birthday Trust. The lower handle, 5in long, has lost its wooden handle. There are six serrated blades, 17in long, that mesh into five plain blades. The longer handle shown only in part, is 21½in long and turned over to form a hook at the end. John E. D. Touche, of Edinburgh, suggested that it was a turnip-slicer, the metal blades being fixed to a post and the long handle, now assembled above the shorter one instead of below it, being worked with a treadle. An elderly visitor to Mary Arden's house has since confirmed that, as a small boy, he used such a turnip-cutter.

The picture below sent by W. R. Mitchell, shows Elizabeth Maudsley, daughter of a Ribblesdale farmer, at a grinder which is fastened to a pillar of stunted oak in the attic of the rambling farmhouse at Stainforth Hall. It is of the type commonly used for 'kibbling' or coarsely grinding beans.

Cutting and tying hay

Ousted by the baler, few hay presses were still in use when these photographs were taken in 1956, yet this example was less than 30 years old. J. A. Smith, of Stocking Pelham, Hertfordshire, who is shown inserting a needle and operating the press, was able to press and truss about 10 tons a week.

Below, top, is a straw twister, often known as a wimble or womble, used by Mr Parradine, a straw-rope maker of Leverstock Green, Hertfordshire. A strap was passed through the loop on the end of the handle and was then fastened round the user's waist to secure it. Before the introduction of binder twine and wire, straw rope was in great demand for tying trusses of hay. Jean Alexander sent the 12in wimble (below, bottom) from Beddgelert; it was used to make bonds to secure thatch on farm buildings. The hook was thrust into the straw or reed; and the tie-maker, walking backwards, kept one hand on the 'toggle', while grasping and turning the body of the tool, as one would a brace for tightening nuts on a car wheel. A second man fed straw or reed into the tightly twisting rope.

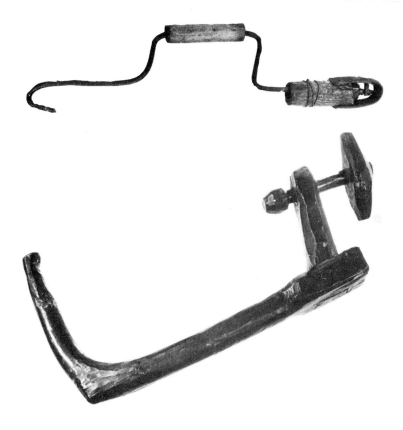

J. Hurst of Denmeed in Hampshire owned the wimble which belonged to his father, a hay trusser, who made his own bonds. If he was alone, he made them on a bond-stick, weighing about as much as a walking-stick and some 18in long; but if someone else was around, he used his wimble. 'The helper had to turn the handle as he walked backwards, while my father fed in the hay or straw from a prepared heap. It took approximately 10 seconds to make one bond for a truss 2ft 2in by 3ft 3in. The measurements were critical: two trusses laid across the wagon made a load 6ft 6in wide, and on the next course these were tied in by three trusses laid down the wagon—6ft 6in again.'

Below are the three 2ft rods of a hay tester with the case above them. They were brought by A. C. Brown, whose father lived near Nottingham and was a member of the Midland branch of the Hay Dealers' Association, now disbanded. The pointed end was inserted in the rick to test its solidity, and a small amount of hay was withdrawn from well inside it by the hook in the head. Similar tools are sometimes used by valuers today.

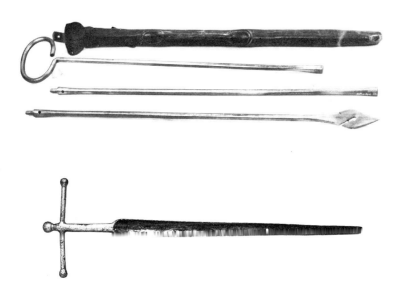

Auger

The T-handled metal auger on the previous page, from Fleet in Hampshire, has a blade 3ft long and, in the lower part, two holes which were possibly for attaching a wider blade. It is too delicate to have been used for cutting into anything very hard, and the form of the handle does not suggest that any great pressure was applied to it. We asked for readers' help in pinpointing the use to which it was put. Several readers thought it was a reamer for enlarging holes in wood, but they usually have much larger handles, such as would be needed to obtain sufficient purchase to turn them in the wood. V. R. Jeans of Epsom, Surrey, believed it to be a soil sampler and has the support of other correspondents, including a head gardener who has tested subsoil with a similar tool, bought about 1880. The two holes in the blade, he explained, were to help marry it to the soil and stop this from sliding out when the tool was withdrawn. He added that in his early days, 'I often had to water a very large peach-house, and when I told the head gardener I had finished, he would make a series of tests with the auger to see that the job had been thoroughly done'. A number of correspondents suggest that the tool illustrated was used for sampling butter or lard from a keg, or a whole cheese. P. A. Hardy of the Canadian Audubon Society, Toronto, had seen a similar tool used to sample cheese in Ontario and had heard of one for taking a sample from a sack of grain. Frank S. Cheney of South Wigston was able to send a sampling spear of very similar design from the Leicester County Weights and Measures Department; it is now in the Museum of English Rural Life. It is only 20in long and has two rivets in the blade to secure cross-pieces, which fill the hollow. It was inserted in the neck of a sack of meal or fertiliser to withdraw a sample, but its use was prohibited by regulation in 1955. The cross-pieces helped to ensure fair samples. It is anybody's guess to which of these many uses the Hampshire auger was put, but it is quite clear that tools of this type were mostly used to take samples.

The Burrell

During the 1939–45 war nearly all the threshing on farms in our part of East Anglia was done by teams under contract, using a Burrell steam traction engine, wrote Richard Williamson. I remember a day in early March when light snow flurries swirled and drifted with the steam and smoke down the village street as we came out of school. The team had come to thresh the last stack of the season, an oat stack at the end of our farm. They had put it off till the end of the winter, for there was a difficult approach—a mile of cart tracks ending in a steep hill through woods, where the going was wet and slippery. The engine took on extra water at the river bridge in the village, and when we had coiled up the hose, green with weed, Bob the driver let two of the children get up into the cockpit for the ride to the farm.

It was before the drying winds of spring, and in places the tracks were channels of water which the engine's seven-foot wheels squeezed out in muddy jets. To us, following the grotesque rumbling caravan was as exciting as the threshing. Out of sight of Bob and his mate Albert, we took rides on the draw-bar of the drum and

elevator, jumping fearfully through the gap between the heavy iron wheels.

At the 'grupps' by the meadow the Burrell took on water again before the final ascent. The driver attached an iron cleg to each rear wheel and then, getting steam up, started off slowly over the first fairly easy gradient. Before the final hill there was a narrow gateway set on a right-angled bend in the track. This he took as fast as possible, swinging out wide into the field, the wheels sinking half a foot into the soft ground. He spun the wheel furiously for a straight approach to the gateway; it had to be right first time, for it was impossible to back and he wanted a little speed for the slope ahead. The gate had been lifted off its hinges, and the gatepost showed the scars of former years. But this time he was through, and the ascent began.

We stood back and were a little awed as we watched black smoke and sparks roar from the funnel. The wheels began to slip on the clay track. When the clegs came round the black monster shuddered as the wheels got a grip and, rearing two feet off the ground, moved

forward a foot before coming down with a thump that made the earth quiver. Gradually the Burrell, slithering from side to side and with the head of steam almost exhaused, heaved itself and its load like a prehistoric monster on the point of extinction and settled against the stack.

Next morning we could hear the drone of the drum a mile away. When the note dropped sharply we knew that Albert had inadvertently let a 'shoof' of oats fall into the drum without cutting the string. This did not happen often, but it always caused amusement, as the whole village knew what the sound meant. When we arrived, the top had been taken off the stack; and drum, elevator, stack and the men's jackets were powdered with a fine snow that came on an east wind. The men were too busy to look up. My father was forking shoofs on to the drum counter, where Albert's knife flicked and flashed as he ripped the binder-twine. At the side of the drum Charlie, an old age pensioner, was bagging up the chaff, his hair made even whiter than usual by the thistledown and dust which blew out and swirled away with the snow.

The Burrell was a magnificent sight threshing. Bob took us up into the cockpit and opened the fire-door for us to warm our hands. Next to us in the engine-well the sleek connecting-rods moved with slight clacking, gliding back and forth with a silken precision that could hold the attention for hours. The machine gave out warmth and power, and a smell of hot oil and steam. The smoke, chuffed through the wire helmet on the funnel, turning from black to white as the coal burned down, and the whole machine shuddered slightly, rocking on wheels which gradually sank into the ground. It drove the drum with a long belt that was twisted to keep it steady. From my view-point I could watch the flap on the join of the belt travel fifty feet away from the engine, flick round the driving wheel of the drum and glide back towards me.

By midday half the stack had gone. The engine turned over gently, as the men gathered round its warmth to eat their lunch. The drum made a low moan as it idled. Now was the chance to look at it closely, for to get anywhere near the chaff outlets when it was running needed goggles. We peered into the drum, at the sifters and riddlers with corn stuck in their grills, at the fans which blew out oat husks through one funnel, charlock and dock seeds (sold for budgerigar food) at another, thistle heads at the bottom; and at the maze of wheels and belts. By the middle of the afternoon they were near the end of the stack, and the rats started to run. Once a weasel popped out and rippled away into the new straw stack under the elevator.

When the last sheaf had been thrown up with a cheer from all the men, and all the rats had gone, Bob got down from his engine and said to my father, 'That's the lot then, Master'; and we helped him untackle. In those days, for a youngster, 'trosh'n' was as good as a visit to the fair.

Part Two
RURAL TRADES AND INDUSTRIES

The Blacksmith

Introduction

For many village children the smithy had all the elements of drama: the dark and sooty interior, the sudden flame, the acrid scents, the hissing of quenched metal, the strength and skill of the smith himself, and sometimes the violence of the horse and the fear of danger.

The growing interest in riding has brought an increased demand for the farrier. He may take his portable forge and light alloy shoes to his customers in the back of a van, but his tools are still the traditional ones. So in this section we look at other tools and especially at a skill which has vanished from the village: the tyring and bushing of a wheel. Blacksmiths all over the country read *The Countryman* and they wrote with enthusiasm to answer our questions. It is from these men that we learn of the suspense of the task, of the controlled speed with which the wooden wheel was tyred with the red-hot iron hoop, and of the mishaps that could occur.

Wheel and branding iron

The barrow-wheel (top right), unusual as to style, is a graceful example of the blacksmith's craft. It measures 19in across, weighs

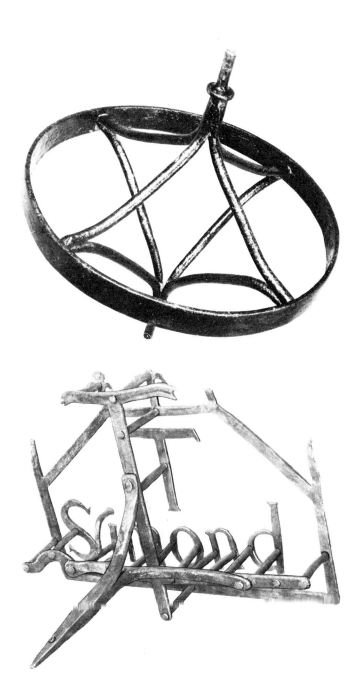

19lb and was made eighty years ago at Methwold, Norfolk, by Evelyn Boldry the blacksmith, for a local builder, Robert Howes. The barrow was still useful to Mr Howes's grandson, who has replaced the iron wheel by a pneumatic one. Another ingenious piece of blacksmith's work is the somewhat elaborate branding iron (below on the previous page), 12½in by 9¼in. It came from Major M. H. Simond's family farm at Binfield in Berkshire. Nobody can remember it in use, but this is the type of tool often used to mark sacks and hop-pockets.

Guillotine and drill

The blacksmith's guillotine was found by H. B. Shields on his Berkeley farm in Gloucestershire when he took over about forty years ago. The point on the shorter handle was fitted into the tool hole of the anvil. The guillotine, which was used for cutting metal,

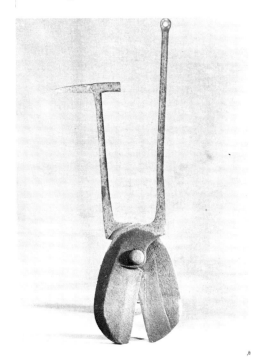

is 36in long. The unusual hand-drill (above) was lying in a black-smith's shop at Weardale, Co Durham. The main stem, 18in long without the bit, was held in the left hand and the crosspiece was pushed up and down with the right. Mrs Anne Peacock wrote from Dereham, Norfolk, to say that her husband was a diamond mounter and used a similar drill, thought it was smaller and lighter. In the jewellery trade it is called an Archimedes drill, and its advantage lies in the fact that it is the only drill which can be used with perfect control in one hand.

Calipers

The calipers (next page, above) were seen by M. Williamson of Hardgate, Castle Douglas, in a local workshop. The owner, who mends lawnmowers and grass-cutters, took over the shop from a joiner who, working with a saw, would have more commonly used a ruler and pencil to mark up work. Calipers are usually associated with the blacksmith, who gives a couple of hammer blows to the iron and checks against them; another blow and check again, and so on until the part is exactly the right size. Calipers without the central 'T' take external measurements, and internal measurements if the arms are crossed to form back-to-back 'Cs'; in this position they can be inserted into the mouth of a pipe or a box. The Hard-

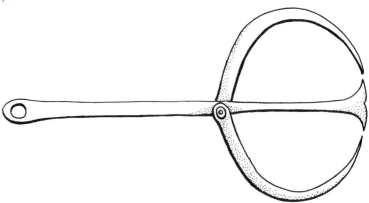

gate tool, 21in long, holds two measurements at once. A black-smith making bolts might use one arm to the 'T' for the length, and the other for the diameter. B. Preece of Nottingham added that he regularly used such calipers in his work at a colliery. 'When forging or drawing down metal to a rectangular section, the arms are set at the two dimensions required. The long handle is to keep one's hand away from the heat of the metal being worked.' He reminds us that blacksmiths' tools are either short and stubby for strength in use on cold metal, or elongated and sometimes mounted on wooden handles for hot work.

Nail-drawer

The claw-like instrument (below) with movable eye came from Sudbury in Suffolk. It is 19$\frac{1}{4}$in long and we tentatively called it a cooper's tool. Pincers will pull a short nail, a claw works well on a large-headed tack, but from all over Britain we heard about the superior efficiency of this tool. The movable eye drops over the nail, which is gripped against the chisel edge as the handle is lowered. With a block of wood under the heel the traction is tremendous, and

the work can be eased by lengthening the lever: a tube slips easily over the tapered end. The household size made for E. C. Robinson's father measures 24in, but S. Tucker used one 32in long in his Dorset forge. The great advantages of the nail-drawer ('devil' in Shropshire) were that it could be used on a nail that had lost its head; and that, by taking short pinches, the longest bolt could be removed from wood or iron in a fit state for re-use. L. F. Kirk wrote that his grandfather used such a tool for removing copper bolts from wooden vessels wrecked on the shores of the Isle of Islay. Until ten years ago it was employed by G. D. Durtnell's firm in Kent—'builders since 1591'—in dealing with half-timbering where old oak formed part of the structure; sawyers used it, too, when extracting metal embedded in a tree. In a West Country foundry the mild steel rods used for reinforcing casting cores are still removed with its help. Tom Grix and F. W. Love called it a linchpin-drawer. The chisel edge eased up the head of the pin, which was pinched out in the same way as a nail. Two readers recognised the tool as part of a wheel jack. G. A. Baker's hung by a chain from the cart and was used only on the wheel: 'To hold this at just the right height and line, and to ease it back on to the axle was tricky. As long as the weight of the wheel was supported by the tool and chain, it was easy'. F. Stanton's jack was used on the axle or hub. 'It is about 50 years since I used one when washing and greasing the tall-wheeled bakers' and butchers' vans', he wrote. 'The base was of cast iron and on the legs was the maker's name—a firm at Netherton, Dudley. I always thought it a mazy and awkward tool. It was difficult to get under a cart and would not fit under the low undercarriage of a baker's van. I mind the best method was to place it outside the wheel and lift by the brass on the hub, then put a box under the axle.' The wheeljacks are illustrated overleaf, above.

Tyring wheels

When George Hogg moved into his cottage at Egremont in Cumberland he found two forges, account books dating back to 1901 and a number of smithing tools. A stone 'wheel', 5ft in diameter, came to light later under brambles, nettles and rubbish, and local farmers

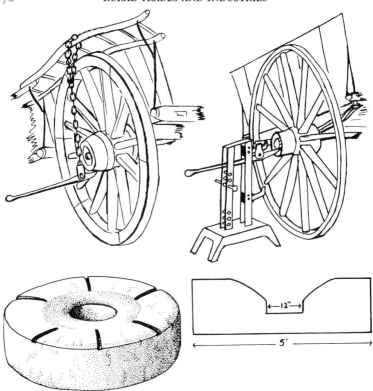

told him that their cart-wheels were tyred on the stone (above). In the accounts George Hogg found '4th Dec. 1901. Cart-wheel rehooped: 2s 6d'. The last entry for rehooping came in 1947, when the price was 10s. Hooping was commonly done on a circular iron platform, the unshod wheel being screwed down at the hub. The hot iron tyre was hammered home over the wooden rim against the platform and then shrunk on by quenching it with buckets of water or by lowering the whole platform into a tank. As stone does not stand extremes of heat and as there was no device to secure the wheel, we were at a loss to understand how the 'wheel' was used. A local man thought it was a 'shoeing hole'. He suggested that the hot tyre of the wheel, slung vertically, was turned through water held in the central depression on the stone; and that the evenly spaced grooves carried away surplus water, while keeping the men standing round the wheel dry-shod. How was the hot heavy wheel hoisted, we wondered.

News reached us of only one similar dual-purpose tyring platform. Also in Cumberland, it is slightly larger and was made in two pieces with a water-tight cement joint. When the wheel was laid on its side to receive the tyre, the hub rested in the central hole; the lighter trap-wheels were screwed down through the hub to a metal ring leaded into the base. The stone lacks the spoke-like runnels but has extensions on either side of the whole, which would steady the wheel, as a rut might, when it was in the upright position for quenching. S. H. Scott wrote: 'Our blacksmith Harold Simpson of Great Blencowe, now in his seventies, is descended from a long line of smiths and once helped to hoop fourteen wheels in one day. The stone for his platform probably came from a quarry, now disused, about half a mile away; the smithy and most of the eighteenth-century buildings in the village are of the same stone. The hooping was done by at least two men. The red-hot hoop was dropped over the wheel and quenched to some extent by water thrown from buckets. Then one man grasped the spokes and lifted the wheel on to its rim, so that the bottom of the tyre rested in water in the central hole. He turned the wheel slowly, holding it by the spokes, while the second man hammered the hoop from either side. Our blacksmith devised an improvement on this method. When the wheel was lifted into the vertical position, a length of steel piping was slipped through the hub, the far end being pushed into a hole in an upright piece of timber. The wheel was slid along the piping until it was over a specially made trough, then revolved in the water to cool as the hoop was hammered into position'. S. R. Hughes, of South Kirkby near Pontefract, reminded us that tyring platforms were not always round 'Square ones were very handy for hammering out large flat objects. The platforms I remember could be anything like 6ft square or in diameter, with a hole for the hub at the centre. The blacksmith did not put white-hot metal to a wheel. He fashioned his tyre, made sure he had something like a good fit, then got plenty of help. When the tyre was what we call just above red-blood heat, it was hammered and slaked at the same time—some hammering, others slaking. The blacksmith had special tongs (below) to pick up the hot tyre. Putting the jaws on the hot steel, he knocked a fastener down the handles; then he did not have to keep a grip, but just to support the weight. Three or four men worked with tongs on the hot tyre, and three or four with the water.'

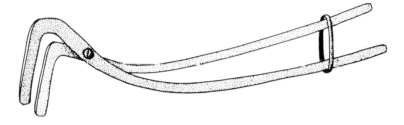

Bryan Young, of Cheam, told us of another solution to the problem of tyring a wheel: 'A two-wheel Canadian cart, popular at the beginning of the nineteenth century, was known as the prairie violin; the squeal of wood rubbing on wood came from the un-greased axles and hubs. In the dry summers the carts moved in a cloud of dust, and dirt would have mixed with the grease to choke the hubs or cause rapid wear. The wheels were made in the traditional way with specially selected woods but, in place of the usual iron hoop, poorer folk bound the rims with strips of wet rawhide, which tightened as it dried'.

Charles Heyte of Paxford, near Chipping Campden was photo-graphed (opposite) using a wheelwright's traveller, to measure an iron tyre for a wagon or cart wheel. On the stump behind is a swage block for bending metal, hot or cold. Sometimes the wheel was shod with iron strakes (opposite right) instead of using one continuous hoop. John Jenkins, who made a special study of the subject at the Museum of English Rural Life, wrote 'Despite the fact that hoop tyres were known in Britain as early as the iron age, straked wheels were common in hilly and clayland districts up to the end of the nineteenth century. In East Anglia and Lincolnshire

they seem to have disappeared in the 1820s, but in Shropshire and Herefordshire they persisted for as long as wagons were made. They were preferred in hilly districts because they slipped less than those with hoop tyres. In clay-land districts, where broad-wheeled vehicles were used, the fitting of a hoop tyre was extremely arduous, and in the Vales of Pewsey and Berkeley, for example, it was customary to shoe wheels with two, if not three, lines of strakes. The tyre of a dished wheel was coned to allow for the conical shape of the wheel, and it was much easier to cone and curve a strake no more than 2ft 6in long than to cone a hoop with a diameter of 6ft or more. The fixing of strakes to the wheel is also a far easier process, requiring less elaborate equipment, so that the whole job could be entrusted to a relatively unskilled man. By fitting strakes the wheel-wright saved fuel: a large number of strakes could be placed in a furnace to heat at the same time, but only two hoops at most. In dry weather the whole wheel shrinks, and strakes, being fitted with nails, stay on, whereas a hoop tyre tends to become loose. When strakes wore out the farmer himself could renew them one at a time.'

Boring the hub

George Sturt, in his book *The Wheelwright's Shop*, tells us that 'boxing' a wheel meant 'wedging into the stock that cast-iron "box" which was to run on the axle; inserting, if you like, the central iron socket into the woodwork. When a wheel, with this "box" fixed into it, was slipped on to the projecting "arm" or axle-end it became part of the cart'. From Atherton in Lancashire, A. Leather sent the drawing of a boxing-engine, round which we sketched the hub (A). A hole was bored through the hub or stock for the 3ft shaft; and dogs (B) were hammered in, to hold it steady. The auger was self-feeding so that, as the handle was turned, a wider hole was cut, the diameter being determined by the setting of the cutting blades (C). The design of this engine would not permit it to open up the

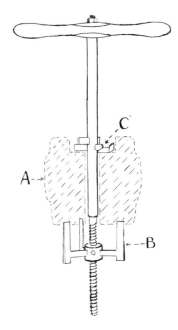

DOGS THREADED

whole socket, and it was used to make a seating for the rim of the box and the linchpin recess. More usual were dual purpose tools such as that in J. E. Skyrme's collection of rural bygones at Dorchester, Dorset. It has adjustable dogs which screw on to the three arms of a round carrier and could be used on hubs with external diameters down to about 4in. The shaft is threaded to within $2\frac{1}{2}$in of the cutter. Cutters of various lengths enabled the machine to be used to bore right through the hub or just to make the seating. From Hardingstone in Northamptonshire, H. Matthews wrote as one of a family of blacksmiths and wheelwrights whose business was started by his grandfather about 80 years ago in the county town. His borer (overleaf) has a second three-legged dog sliding along the centre shaft and fixing into the hub at the opposite end to the screwed one. 'First a hole large enough to take the centre shaft is made through the hub', he explains. 'The dogs are then fitted and hammered into each end, so that the shaft runs through the centre. A short cutter is inserted and fixed by a grub screw. After the main hole has been cut to size, each end of the hub is turned out with a large cutter to take the enlarged rear end of the axle-box and, in front, the cap, collet and pin.' James T. Robson, who represents the third generation of cartwrights at Matfen, Northumberland, told us that he still had in his shop a 'bushing-machine', as it is called thereabouts. The design is again different, the nave being held firmly by set screws in a cradle with iron hoops approximately 15in in diameter. By adjusting the screws the auger is brought exactly to the middle of the nave. The handle could be used on either end, and the boring done from both faces of the knave. The cutter was also reversible. The hole was bored a tight fit for the bush, which was driven in with a 7lb hammer, but a little room was left at the flange end so that the bush could be centred with wedges.

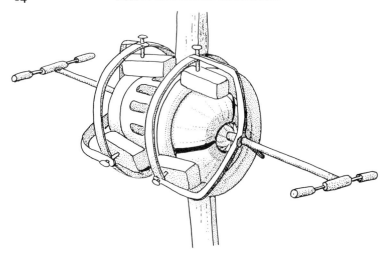

Spoke dog

Below is a spoke dog, used to strain together two spokes of a wheel, so that the felloe might be tapped on. The iron hook was placed over one spoke and the bottom of the 3ft ash handle behind the other. By pressing forward on the handle the two spokes were brought together and the felloe was then slipped on.

'Fork' with flattened tines

The 'fork' illustrated above was found by John L. Gilbert at Wansford, Northants, in the outbuilding of an old house long occupied by a lady who used it in the garden. A farm worker, thought to have been a horseman, lived in the house before her, so the tool may have had some agricultural use. The iron part is $10\frac{1}{4}$in, and the handle $4\frac{1}{4}$in long. W. R. Harding, of Ancaster, Ontario, told us that tools much like the 'fork' are commonly used in modern industry for removing shock links, tie rods, pulling gears, bearings, etc. 'The flattened tines', he goes on, 'are necessary so that they may be inserted behind the pieces to be removed. While I am not knowledgeable in the art of wagon making, I believe there were several pin joints which would require the occasional application of grease; perhaps this tool was used to pry out the pins'. The old lady, we discovered, who used the tool in her garden had relatives who were wheelwrights.

Grease-box

The grease-box below is about $2\frac{1}{4}$in deep. It was used in a workshop (now a room in the house) which, with the bedroom over it, used to

be in Wansford, Northamptonshire, divided by the county bound-
ary from the rest of the building, which is in Stibbington, Huntingdonshire. Such boxes, containing animal fat, were to be found in
every wheelwright's and carpenter's shop before the days of oil-
cans. In this shop grease was also kept in a hole measuring 6in by
2in by 1in in an old oak beam.

Coachbuilder's clip wrench

We asked for readers' help in identifying the 11in iron tool for
A. H. Scarrott of Westbury on Trym. The projection or stud on the
upper arm is solid, and there are no hammer marks on the back.
Raphael Salaman gave us the answer: 'It is a coachbuilder's clip
wrench, used in assembling the leaf springs and in attaching them
to the axle of coach or cart. The legs of the clip holding all together
tend to spread and have to be eased through the holes in the plate
to which they are bolted. The stud on the moving arm is fitted into
the bolt hole, while the concave end of the wrench pushes the leg of
the clip into the correct position.' We hope the sketch (opposite)
makes all clear.

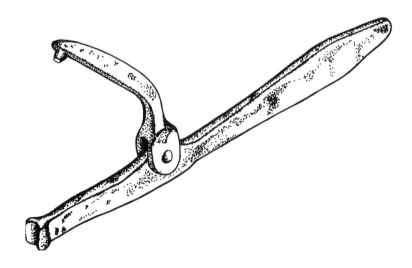

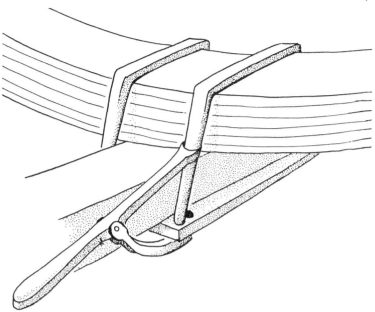

Place the waggon

We asked James Arnold, artist and author of *The Farm Waggons of England and Wales* (John Baker, 1969), to comment on the photograph below.

You might think that all the waggons of any one country were of more or less the same design. It has taken me thirty years of travelling and studying in the field, spotting waggons behind barns, sketching and noting, to understand the combination of features of each design, and to realise that there were affinities which make it possible to group them loosely, at least regionally: East Anglia, East Midlands, the Marches, Wessex and so on, with transitional types between nearly all adjacent regions. One waggon alone will rarely tell much; it may not be a 'native' but from some other county. When you have examined a dozen or more which are similar, you have a good indication. By the time you have seen fifty or a hundred, you can not only establish the county design within each regional group but also note variations on the theme.

I confess to difficulty in Gloucestershire (excluding the Cotswolds) where nearly every wheelwright seemingly went his own way. For all that, there are group affinities. Some of the waggons I have sketched between Gloucester and Bristol had shallow bodies with bow-rails over the hind wheels and large near-vertical ladders, all in the Wiltshire manner. To create uncertainty, others bore 'Glamorgan' half-moon panels on the headboards. Between Gloucester, the Forest of Dean and Tewkesbury the makers appear to have turned their backs on Wiltshire, yet conceded little to the Cotswolds, Hereford or South Wales. Nearly all their waggons ran on broad wheels; yet, in spite of the deeper bodies and absence of bow-rails, they had a kinship with those across the Severn. Whereas most counties had each a traditional colour, Gloucestershire had several—blue, ochre, yellow and salmon.

You rarely encounter a waggon so far a wreck that there is no feature by which it may be identified. A wheel, part of the forecarriage, or the panelling may produce the clue. I believe the waggon photographed by G. D. Bates may have been built north of Gloucester. Firstly, the broad wheel is tyred with coned strakes on the outside of the rim and a flat hoop on the inside—a practice exclusive to the county and the adjacent part of Wiltshire. Secondly, iron chafing-plates are fitted to the hind-wheel spokes where they enter the felloe-ring or rim, to take the wear from tie-chains wrapped

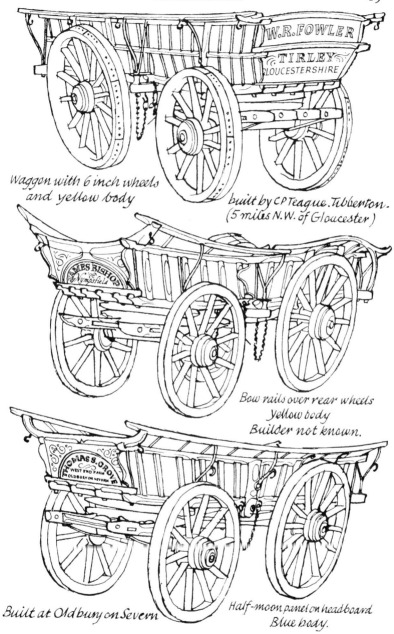

W.R.FOWLER
TIRLEY
GLOUCESTERSHIRE

Waggon with 6 inch wheels
and yellow body

built by C P Teague. Tibberton.
(5 miles N.W. of Gloucester)

JAMES BISHOP
Nympsfield

Bow rails over rear wheels
yellow body
Builder not known.

THOMAS S. GROVE
WEST END FARM
OLDBURY ON SEVERN

Built at Oldbury on Severn

Half-moon panel on headboard
Blue body.

round for braking—again a feature common in Gloucestershire, though also as far north as Bishop's Castle in Shropshire. Thirdly, every waggon has the sides supported midway and at the rear by standards of wood or iron. Either material was used in Gloucestershire, but this waggon is exceptional in having the Hereford wooden 'elbow', which in turn is unusual in being secured by a vertical bolt through the cross-beam instead of the usual wedged end fitted into a mortice. Fourthly, in place of the customary wooden cross-piece at the front of the body-frame there is a long iron strip—quite usual in the county.

The moderately deep body, the apparent absence of bow-rails, the lightly built forecarriage and the waisted bed that gives the wheels greater lock, taken separately, are not conclusive; but the broad wheels with strakes and hoops, and the Hereford elbow standards indicate a builder north of Gloucester. I note the number of very thin spindles along the sides without coming to any conclusion, and am most of all puzzled by the large disk on the head-panel. Such a motif has been found only between Kettering and the Wash. G. D. Bates, who found the waggon he photographed, states that it 'did not appear to have any pigmented colour; the under-carriage, which was pink, looked as if it were bleached'. By the time waggons reach this condition, they have often lost all colour and are a mass of fungi, moss and mildew.

We must beware of false evidence. In a museum I found a waggon model described as West Gloucestershire. It was almost identical with those built by Bradley, of Llanfihangel-ystern-Llewern, and the original came from Llanfetherine near Abergavenny. Some time ago I examined diagrams of a waggon built at Bodiam and described as Sussex; though built by a wright of that county, the design was from Kent. At least one man in West Sussex made waggons similar to those built by Sturt at Farnham in Surrey. And here are three examples of waggons which travelled far from their native heaths. In a field at Michaelchurch-Escley, where Herefordshire runs over the Black Mountains, there were massive deep-bodied Suffolk waggon and a half-open-sided waggon, both from Wymondham. I wonder how they negotiated the steep

Broad-wheel with strakes on the
front ring and hoop on the back
GLOUCESTERSHIRE   Broad-wheel
with 6 inch strakes having
diagonal joints   HEREFORDSHIRE
A wooden 'elbow' supports the side of the body

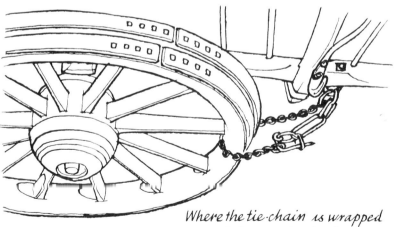

Where the tie-chain is wrapped
round the wheel, chafing
cleats are often fitted
GLOUCESTERSHIRE

and mountainy tracks so different from their own broad fields. At Caynham, south-east of Ludlow, I found a late Devon waggon, small and light, built by Milford of Thorverton, together with a two-wheeled Welsh gambo built at Sennibridge. And at Goring, below Oxford, I found a plain straightforward Kent waggon and a graceful ship-like Oxford side by side.

It is not easy to see why the designs developed a geographical pattern. Gloucestershire has a dozen, Oxfordshire but one. This is found, with minor variants, in large numbers everywhere between Winchcombe and Windsor, and from Inkpen Beacon to well north of Banbury—through seven counties. Small wonder that there is difficulty in finding a generic name for the group. I confess an affection for these waggons. They looked and were efficient. They weighed light, ran well and, with no more than two horses, carried their loads easily; and they had a pretty good lock. The majority had 'crooked beds' with divided frames overlapped to give a shallow insection, though most of those built south of the Thames had 'waist-beds' giving a deeper insection. A few in Stamford Vale had broad wheels, but nearly every other waggon in the region ran on large hooped wheels only $2\frac{1}{2}$in wide; the diameter was 50in for the front and 60–62in for the back.

If the Oxford was successful from Cirencester to Banbury, why not between Banbury and Stamford? In Lincolnshire there was no resemblance between the decorated waggons of Holland and the simple robust lines of Kesteven-Lindsey. If the majority of Shropshire waggons found between Ludlow and Whitchurch were plain and blue, why were the few west of Craven Arms elaborate and yellow? I fancy the answer is probably sociological rather than geological or agricultural.

It seems that the wheelwrights themselves hardly knew their waggons by their counties. Their day-books contain entries such as harvest waggon, narrow- or broad-wheeled, 6in waggon and so on. In other words, they were known by the width of wheel rather than by the county. The waggon in the photograph found its last resting place at Hucclecote, two and a half miles south-east of Gloucester.

Building

Introduction

As one travels across the country it is still possible to see by the buildings where clay replaces limestone, and where wood was once abundant. Even in west Oxfordshire, when the wolds were more thickly wooded, cottages were built with a timber frame and thatched; for at the time these less durable materials must have been freely available. Small domestic and farm buildings in particular reflect the cheapest materials of the time and place.

Regional materials are of course reflected in regional tools. It is to be expected that a man working on brittle blue slate from Wales would use tools that were more precise than those used by a man shaping tough limestone.

Combers

The large reed comber (overleaf) made by J. Blackler of Kingsbridge in Devon, may have been powered manually; but the pulley, handle or fly-wheel is missing, so it is difficult to be sure. A Cullompton farmer recalls a similar machine driven by an oil-engine. To prepare the straw for thatching, it was fed by hand between the long inter-meshing teeth set in the faces of the two cylinders, which revolve towards each other, and held firmly against the teasing action of the teeth. It emerged on the other side to lie between two iron arms (shown here in the closed position). When there was enough straw for a bundle or yealm, the two arms were cranked together to compress the straw, which could be tied by hand. The distance between the arms and drums can be altered to take long or short straw or reed. On the much smaller and simpler comber (opposite) the bundle of straw was laid on the teeth and pulled off as often as was necessary to work out the tangles and get the straw lying neatly, all in one direction. The Somerset man (opposite below) photographed in the 1940s, was pulling horse hair over a comb, but the method is the same.

Thatching needles

Margaret Marris of Blackhurst Farm, Rushlake Green, asked us to confirm her identification of the 22in thatching needle below. Thatch is held by two sets of ties: the binders run parallel to the ridge over each course of thatch; the second set stitch the binders and thatch to the framework of the roof—the battens and rafters. The binder may be of split or round hazel (sometimes used in a decorative finish), and the 'stitch' may be an iron hook driven into a rafter, although quite frequently the thatcher binds and stitches with tarred twine. In earlier days a boy was stationed in the roof space to pull and push the long straight needle through the thatch; but the more complicated needle below enabled the thatcher to work single-handed. The curve of the tool fitted round the bundle of straw, and the eye in the tip of the needle carried the cord through when the plunger was pushed home. The illustration shows the tool in the home position.

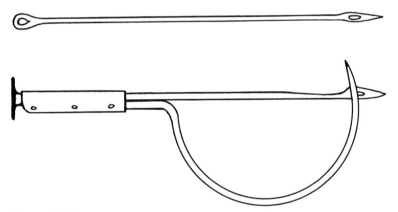

Bramble hook

Bramble stems were sometimes used to tie the thatch to the rafters, or to bind the straw coils of a bee skep. Above right, at little less than actual size, is an iron briar or bramble hook with steel tip; the long handle is missing. It belonged to a champion maker of bee skeps, Job Swinford, of Filkins, Oxfordshire, who used it to cut the

brambles with which he bound the wheat straw. After cutting them, he pulled them through a ring to remove the thorns, split them on a blade fixed in a table, then bundled and dried them. Mrs R. Rawbone, now living at St Leonard's-on-Sea, describes for us her father's method of making bee skeps at Brize Norton, only a few miles from Filkins. He used a strong pocket-knife for cutting, stripping and quartering briars, and for sharpening the ends for sewing. He began the skep by pulling a handful of straw from a bolton and passing it through a cuff, which kept it together and at the same thickness throughout. The cuff was 6in long and the size of a man's wrist, and it was made of a strong piece of hand-sewn leather.

Slater's tools

In the debris of a 300-year-old barn which was being demolished to make way for houses at Flixton near Manchester, T. McLachlan picked up a peg $2\frac{1}{2}$in long and made, apparently, of oak. It had been roughly tapered and driven tight into the hole of a stone slate, known locally as a 'tun' slate. This was hung by the peg from a lath laid across rafters. When a roof begins to dip and slip, it is usually the laths, not the pegs or slates, that have perished. The Cotswold slater did not bother to taper his pegs. Working with harder stone, he used a smaller hole and drove the peg home with a wooden mallet. The wood 'reeved' up, or swelled round the constricting stone, to form a very adequate 'head'. George Swinford of Filkins, near Burford, used the splitter illustrated to make pegs, preferably from yew and the hearts of larch. After he had tied string round a block 4in by 4in, he laid the splitter about $\frac{1}{3}$in from the edge and gave a sharp tap with a hammer. Having worked across the block,

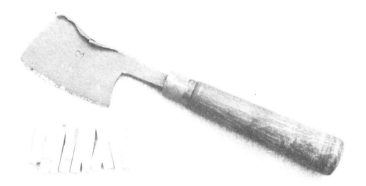

he gave it a half turn and split it in the other direction. When he untied the string, the block fell apart into a gross of pegs.

Readers reminded us that stone slates were not always hung by wooden pegs; and Dr M. L. Ryder refers to the parish accounts from Pannal, Harrogate, where it is recorded that in 1678 Henry Winterbourne was paid 17s for slating the roof and 1s 3d for 500 sheepshanks. Writing from the office of the Royal Commission on Historical Monuments (England), R. W. McDowall recalls a roof in South Lancashire where all the pegs were of bone. This seems to have been used more in the north than in the sheep-walk country farther south, but 30 years ago J. W. Stevens saw an old building on the road between Cheltenham and Stow-on-the-Wold pegged with sheep's ribs. It would have been difficult to make a small hole and drive a wooden peg tightly home in soft stone—it was sometimes necessary to drill the hole—and the light bone with its natural head would have solved the problem.

The hole in the hard Cotswold slate was made by carefully slanted blows with a slat-pick or pittaway, and the hole of a finished slate was surrounded by peck-marks. The head of the tool was set at an acute angle to the handle to accentuate the glancing element in the blow. It was often made from an old file sharpened at both ends; a quick tap of the securing wedge, and it was only a second's work to reverse the head. A man might carry a dozen heads with him, for the points soon blunted (opposite below).

Welsh slates required different treatment: one sharp blow for each clean hole. C. B. Dicksee, now living at Seaford in Sussex, remembers men making these slates on a building site in north-west London at the turn of the century with a curved and spiked tool similar to the straight-edged one below. The slater sat at one end of a trestle and worked over a raised iron bar which ran along it. After scribing the slate with the spike, using a wooden rule as a guide, 'with his left hand he dropped the slate on to the bar with the line immediately over it and chopped with the blade. Then he moved the slate across a few inches and punched two nail holes, each with a single blow, using the spike on the back of the tool.'

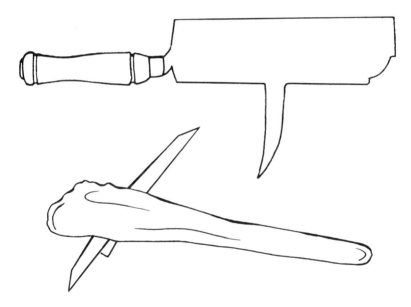

From the north Cotswolds has come another stone-working tool, the crapping or slate iron (overleaf above), on which the slatter laid the thin stone flat to shear or trim it with his slat hammer. This specimen is 11in high and 18in long. At Stonesfield in Oxfordshire a narrow stone called a crapping stone was used for the same purpose, held upright between the knees. In the Burford district the slate itself was set on end for trimming.

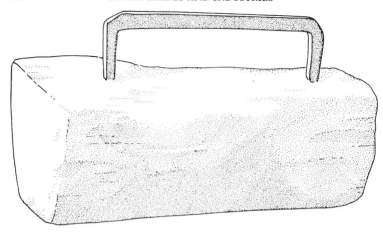

Thwacker

In the mid-nineteenth century the Duke of Manchester owned a flourishing brickworks at Kimbolton in Huntingdonshire, where tiles also were made. Hip or valley tiles, when half dried, were placed on the dresser and hit with the 'thwacker' to correct any warping during drying. The dresser is 15in long.

Brickmaker's mule

The 'mule', an old tool of the brick-making industry, was hanging on the wall outside the ironmonger's shop in a Hampshire village when Peter Rosser saw it. He photographed it and went inside. There a wiry little assistant told him about it and his eighteen years as a brick-maker between the wars.

'It's a cutting-iron, that is, sir,' he said, 'for chopping up an' working the clay like. It ain't a spade an' it ain't a fork, now is it? "Mules" they call 'em. That's right now, i'n't it? Mules, see. Still got mine at 'ome. Ain't so wore as that 'un, but I got through a lot o' mules in eighteen year, eh? Them grooves like is where you 'old it when you cut an' twist, see. You never got no blisters, mind; your 'ands was a sight 'arder than any ash wood.

'You worked a gang o' four, see; one to dig an' three to barrow away like. There were about twenty feet o' topsoil before you got to your clay, but you didn't get nothing for shifting the topsoil—only clay they wanted, see, for the bricks. We got the clay out o' the pit an' barrowed it into the brickyard, about three 'undred yards, dig to tip. We ran a 'undred yards each like, then the next took over. That meant one man was digging wages for four, see; for topsoil nothing, but a shilling a yard for the clay between us.

'Five in the morning we'd start, year round; but you 'ad three hours' work getting ready to earn, if you follow me, 'cause o' the topsoil. Started cutting the clay around eight or 'alf past, an' we reckoned to stop off about four-thirty to get readied-up in the brick-yard an' get all set for tomorrow like. That was another two an' a 'alf hours, in our own time like; they didn't pay us to get ready, only just for the clay, see. That made it seven at night, with a quarter of an hour off middle day. After we'd stacked it, say eight feet by forty, it were supposed to lie a twelvemonth to ripen for the brick-making, but that didn't often 'appen, with the demand. I

could make about 1250 bricks a day; that was at ten shilling a
thousand, all piece-work like. Best 'and-faced, ten shilling a
thousand between us. Come firing time we got a shilling a thousand
stacking the kiln, but with the 'andling an' barrowing you touched
more like three thousand for one an' six—the four of us, that is.
When I was on firing 'twas five to midnight first day, with the
middle day break, then 'ome for three hours' sleep, then back on the
job, forty-eight hours continuous. Two quid, that. If a lorry come
for a load, you'd get sixpence a thousand for loading it, but that
weren't gang work; you could keep that yourself. Best 'and-faced
ran £4 a thousand then. I took 'ome a good three quid a week—not
too bad in them days.

'The war finished 'em. Ours was all what we called open kilns;
they give a continuous flare like, so they closed us straight away.
Eighteen years I were there. Took over from my grandfather.
Course 'twas real rough in 'is day. There's other firms round about
'oo've bin on to me just to make brick-moulds at fifty bob a thousand, an' none o' the running about. But the wife says no, eighteen
years was enough; you done enough trotting by now. Folk laugh at
me an' say I'm always trotting 'ere, there an' everywhere. It's true,
you know, I just got into the way of it. I don't notice it myself.
We was 'appy, mind. Oh yes, we was 'appy'.

Leading iron

The leading iron (above) was last used in 1905 for melting strips
when they were fitted on windows. It is 17in long.

Paint mullers

From W. A. J. Prevost of Moffat came the photograph of two stone objects above which were not identified, though they were on show for a year at an Edinburgh museum. That on the right appears to have been shaped only on the underside, which has a smooth and true surface; that on the left has been more extensively worked, possibly to a shape convenient for grasping with the hand. Several readers identified them as paint mullers. A. H. Frost wrote: 'My father was apprenticed as a painter and plumber 70 years ago and remembers how in winter the staff prepared pigments for use during the following spring and summer. A pigment was mixed with oil or water and spread on a flat stone. The muller was worked in a circular motion with the left hand, and the paste of pigment was scraped up and kept in the grinding area by a palette knife held in the other. The muller weighed several pounds. The stone and muller were replaced by the iron hand-driven paint mill'. D. Lea of Altrincham described the process in similar terms except that in his recollection, the muller was grasped with both hands. As a lad, J. W. Faulkner of Rugby used to be sent out for some umber-in-oil, which the local decorator would grind while he waited. More than 60 years ago N. Teulon Porter worked many hours with a muller in a wagon builder's shop in the West Riding, and he tells us of two specimens in the Shaftesbury museum: one is of marble and was probably used by an artist for specially fine work. Seventy years ago, as a bookbinder's apprentice, E. J. Watts used a muller on a flat piece of marble to grind the colours for marbled or sprinkled edges for books. Colonel Hosie of Co Kildare had a complete set brought by his great-grandfather from Scotland. His largest muller weighs about 10lb and was used on a smooth flag 2ft 6in square. As a boy he watched the estate carpenter grind paint by this method. The owner of the stones illustrated told us that they came from a carpenter's shop, and finds that the base of one has a thick coating of various pigments.

Wooden guttering and pipes

When she moved to Wallington in Surrey, Mrs C. O'Hea found lengths of wood which she thought to be guttering. Edward Pinto, expert on treen, was able to confirm her identification. Such guttering was in common use in country districts until the early years of this century, though little is left today. A length from Wallington went to Oxhey Woods House. In the early years of this century F. R. Jolley's parents were living at Swansea in a house which had eaves, gutters and rain-water pipes of cast iron on the front elevation; but at the back, facing the sea, they were of wood. The gutter was V-shaped with a small triangular piece of wood in the angle to seal the joint. The down-pipe was, in section, 'a neat little oblong box'.

A. C. Shill of Rayne, Essex, sent the photograph (above right) of the tool for boring water pipes. He found it in the yard of Brock Brothers, builders at Rayne, who estimate that it has been in the family for two hundred years. It is unusual to find so complete a set of gouges. Before the advent of earthenware and metal pipes, wooden ones were commonly used for wells, for conveying water along the ground and even for drainage. These wooden pipes are

often dug up today, and it appears that elm was the most suitable for the job. The boring of them was a difficult task calling for great skill and patience. A small hole would first be made, and the diameter then enlarged, the extension piece being added to the gouges as the tool went into the pipe. The most skilled part of the operation was that of ensuring a true joint between holes started at each end of the pipe, and there are many stories about pipe borers who after a night out failed to make ends meet.

Timber and Barking

Introduction

One of the pleasantest expeditions made for this feature was to Bucknell in Shropshire. We had been sent some remarkable photographs, taken at the turn of the century, and set out in search of more. The man who had taken them was clearly interested in crafts and country industries, for his shots were angled to give as much information as possible. We discovered that the photographer, Ted Picken, was a countryman of remarkable versatility —tree feller, furniture maker and carver, cobbler, garage owner and engineer. The portrait of a thatcher—his knee pads, spars and smock, the ladder with rounded side rails, all in clear detail—is typical of his work.

Marker

The 7in wooden-handled tool (right) with spike and oddly shaped cutting blade is a tree-marker. The forester used it to blaze the bark of trees that were to be felled. From the Boat of Garten, Invernessshire, J. I. Johnson wrote to tell us that the 'marker for blazing trees' is known as a timber-marking scribe in that part of the world.

'It was used to incise identification marks on large logs or baulks of barked timber. Straight lines were made with the gouge alone, circles and semi-circles by rotating it about the spike. By combining these movements one could form every letter of the alphabet, all numerals and many more symbols, to denote ownership, quality and destination. Foresters still use a similar implement, without a spike, to mark the bark of standing trees with a straight scratch or combination of scratches to indicate that it had been counted or measured. The mark, like the implement, is known as a "screeve".' R. Nicholson of Steeple Aston in Oxfordshire uses a modern steel scribe with cranked handle. Holding it as a dagger, he makes bold sweeping down-strokes to number trees. From R. F. S. Starr of Alexandria, Virginia, we learn that in American usage the tool is a 'race knife'. He was given one by a man who emigrated from England as a child with his father and uncle, both carpenters, and who explained that it was used to scribe hewing lines on logs and timber. Many village carpenters and wheelwrights used to fell their own trees, so the tool would have been doubly useful.

Barking tools

Oak bark is rich in the tannic acid used in the preparation of leather. The felled trunk was ringed at intervals of 24in, and cuts were then made from ring to ring. The bark was levered off with the irons in semi-cylindrical plates that could be conveniently stacked. The 27in tool with 'T' handle (overleaf) is a common form of barking iron. The two smaller irons were used to bark the branches or 'wrongs', hence the name wrong irons. The naturally forked handles gave just the right grip. The two bottom irons were discovered about 3ft down near a Stamford building where they

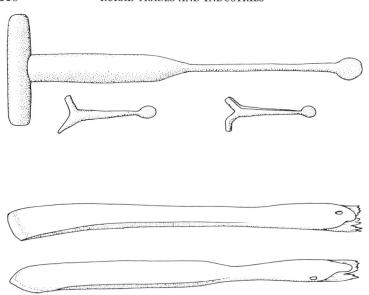

had probably lain for two or three hundred years. They bear certain similarities to barking irons known to the Museum of English Rural Life, though no-one there had seen anything quite like them.

We received confirmation that the Stamford tools were used for peeling bark. J. O. Evans lives in Bucknell, a Shropshire village set among hills of mixed hardwood which supported a thriving industry in bark from 1860 to 1914. 'The fallen oak was ringed with axes in lengths of 2 to 3ft,' he writes, 'and the bark, peeled away with implements similar to those illustrated, was stacked on edge against the felled tree. It was left to dry for two to three weeks; otherwise it became mildewed and was spoilt for tanning purposes. Special containers were fastened to timber carriages to haul the bark to the yards, where it was ricked for six months. The building of these enormous ricks was a leisurely business, as the bark was carried up in wiskets—flat woven lath baskets made locally from riven oak saplings. In the early stages the men went up a catwalk of planks with the wiskets balanced on their heads; as the rick grew they had to climb thirty-rung ladders. It was a skilled job to build and top up a weather-tight rick, finished off by a layer of bark along the crest or ridge. When the bark was ready it was taken out of the ricks, crushed, bagged and despatched by rail to the tanneries at Castleton. The crusher, hand-operated, was similar to a root chopper. The bark was fed into a container to fall on a set of fixed knives; with the turning of the handle a second set worked against them, scissor-fashion, and the chopped bark fell through into the sack hooked on below. In the late 1860s the leather-dressing firm of Ormerod's, which celebrated its centenary in 1968, was buying bark over a 20-mile radius and railing it to Bucknell for ricking, the purity of the atmosphere here being particularly suited to the maturing process. Mr Robert Ormerod, now 92 years of age, tells that the firm was at one time buying 250 to 300 tons of bark a year for the tanning of a thousand dozen sheepskins a week. But the barking industry in Bucknell came to an end as a result of the 1914–18 war, when heavy demands were made on leather stocks. Tanning by oak bark takes from 15 to 18 months and, to cut down the time, the tanners experimented with nuts, foreign barks and, later, chemicals. The time has now been reduced to 3 to 6 months, but (and I speak as a cobbler) very much to the detriment of the wearing qualities of the leather'. The old photograph from Bucknell shows a bark rick in the making, the catwalk and the wiskets.

Baskets of oak laths

The wisket has other names: sometimes it is called a spale-basket; to B. R. Townsend of Wakefield, who sent the drawing of a swiller's horse, it is a swill. This type of oval basket is found mainly in north Lancashire, the Lake District and south-west Scotland. It is made of strips of oak on an oval frame, which is usually of hazel. Oak saplings are boiled in water, split into strips about $\frac{1}{8}$in thick and tapered at the ends. The ends are then secured by wrapping them round the frame, alternate strips being plaited in, when the wood is wet and flexible. To handle the shape and the strips the swiller uses a horse. He sits on the seat and, with his foot, presses down the arm (A), which is pivoted (at B); the head (C) holds the strip firmly against the top platform while it is being shaped with a spokeshave. The horse illustrated was found in an old barn at Finsthwaite, near Newby Bridge, where it had been used by several generations of craftsmen. The swill, usually measuring about 3ft by 2ft, will stand up to a great deal of hard wear; Mr Townend was using in his garden two he had bought twenty years previously.

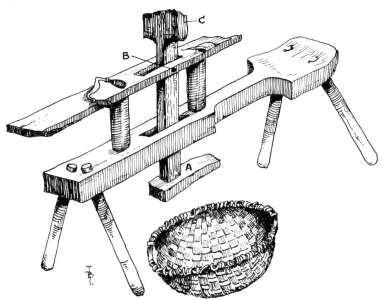

Gearing in a tannery

The photograph of gearing in an old tannery came from Commander E. H. D. Williams of Bath, who wrote: 'In these days of mammoth engineering undertakings it is refreshing to discover the remains of country industries which grew up to meet local needs or utilise local materials. At Holford in Somerset there is a group of buildings with a water-wheel as high as the adjoining house. The small but constant stream which cascades down the combe was thereby harnessed to provide power for a tannery and, in later years, electricity for the house. The wheel's slow motion was geared up through a train of gears, the teeth of alternate wheels being of wood —probably the same oak that provided the tan. The wooden tooth, set in an iron rim and held in place by an iron peg driven through its shank on the inside of the rim, took the wear and was easily renewed.'

My apprenticeship

The craftsman peered up the bole of the elm. 'Ay, farmer, I'll take it.' The tree was at Elm Tree Farm, Heathhouse, on a knoll overlooking the flat Somerset moors, intersected by rhines, where every year the Brue flooded on its slow course to the sea at High-bridge. Thomas Edney, the craftsman, beckoned to me, his fifteen-year-old apprentice: 'See, William? The tree is good—no knots, straight boughs. We'll have it down now.'

William Gooding wrote: Those were peaceful country years. The second world war was far enough off to be discounted. About us were the heavy dairy farms with substantial Georgian houses. The villages of Mark, Blackford and Wedmore lay like the spokes of a wheel; and the bluish Mendip ridge, with the great Cheddar Gorge, formed a low protecting spine behind us. While other wheelwrights and smiths were installing steam-driven machinery and electricity, Thomas Edney's shop remained an obstinate oasis of hand-craftsmanship. The smithy and workshop, a long span-roofed building, stood only a short distance from the farm. Its double doors, open to all winds and weathers, were like huge abstract canvasses, from the cleanings and primings of many brushes. 'Hurry up, boy! Get this lot clean!' the old man would command. Completely content, I would scrub whorls, noughts, crosses, zigzags and cartoon faces until the brushes were dry of paint.

At the shop the elm trunk was hand-sawn into rough blocks. Holes were then drilled through the centres to allow air to pass into the heart of the wood, to prevent radial cracking. It was my job to pile the blocks in a quiet corner on intersecting slats to dry out for a year. Meanwhile we got on with the turning of matured blocks into wheel-hubs. Every day I scrambled up a rickety ladder to the half-platform in the rafters which housed the six-foot wooden wheel connected by a belt to an ancient lathe below. To a chorus of 'Keep it going, boy. Don't go to sleep up there,' Thomas Edney deftly turned the blocks.

The work was mostly seasonal and, as in any other trade, we kept ahead of demand. 'Floods 'll be up soon, boy. We'll get half a dozen flat-bottomed boats ready.' These, at £5 apiece, were made from green pliable elm. I would sit sorting out properly angled wood for the ribs. When assembled, the boat was turned keel uppermost and, amid smoke and anvil sparks, we applied pitch to every joint. Later I would have the job of floating the boat out on to the water. Sometimes, too, there was the pleasure of lying in the bottom to watch the wild geese go over and, on rare occasions, a pot shot at wild duck with a muzzle-loader.

During the winter we did a great deal of tree-felling. Then the midday meal consisted of a whole loaf of bread, a pound of Cheddar cheese, hacked by Thomas Edney with far less than his usual skill from the body of a truckle, and often a whole shoulder of cold mutton, all washed down with cold tea, as we sat on a fallen tree trunk. On wet or quiet days I turned the old paint-mill at the back of the shop, pouring red lead, turpentine and linseed oil into it and grinding them smooth. Or, for hours, I would sandpaper panels of the kitchen dressers for which we were noted, until I felt I could truthfully answer the inevitable question: 'Like silk is it, boy ? Like silk ?' Sometimes I would be set to making iron-ringed hammers of apple wood, popular in the district for smashing coal and peat turves. And the arrival of the eels in the river coincided with the finishing off of the ash handles for five-pronged eels spears made by the smith.

Summer brought a high tide of activity. Rakes, loaders and hand-made wains had to be ready for haymaking. There were cheese presses and butter churns for the milk flush, and apple mills for the cider harvest. The wains, with traditionally blue bodies and red wheels, were made of three woods: oak for the frame, elm for the bodywork and tough ash for the curving front panel. From the local inns and cider-making farms the cry would go up: 'The pummace isn't coming'. Often this meant an enjoyable time away from the shop, sharing the excitement of cider making, while presses were being repaired. For me there was usually a lot of waiting about, and I would be told, 'Help yourself'. Then, sitting on a pile of straw, I would insert a long wheat straw into the trough and, in the warmth and goldness of the day, syphon up sweet unfermented apple juice.

Chief joy of the lunch-hour break was a visit to Mrs Tincknell's shop down the road. Its leaded bow window was almost obliterated by the huge jar of bees' wine for sale at a penny a glass. She had a kind heart for apprentices and schoolboys; and I would return without a care in the world, soothed for the price of threepence by an outsize apple dumpling from the shop oven and two tumblers of alcoholic bees' wine.

Occasionally Thomas Edney went to Bridgwater. He would return with fresh tools, a new leather hide for wheel-washers and half a dozen kegs of paint powder; and his white hair and beard, grown so patriarchal since the previous trip, would have been severely pruned. Every seventh day he laid aside his white carpenter's apron, the ancient paint-decorated hat and the gaming coat with massive pockets. Putting on a black suit and gold chain, he went to the near-by chapel where he was the local preacher. Under pain of royal displeasure I also attended three times on Sunday; and each year I received a book inscribed in the familiar copper-plate: 'For regular attendance at Heathhouse Sunday school'. On Monday morning I was asked: 'Did you enjoy my sermon?' 'Yes, sir', I would reply, upon which we would settle to work. Thomas Edney was an inveterate hummer and whistler of hymn tunes. I would find myself swinging the plane in time to the slow cadence of 'Jesu, Lover of My Soul', only to be shaken by its sharp cessation and the command: 'Smarten up, boy! You'll never be finished'. 'If only he would try "Onward Christian Soldiers" or "Fight the Good Fight",' I complained to the smith, 'I'd get on better'.

When the fifth spring arrived, my apprenticeship years were up. Looking over the moors, I felt I could stay for ever. The floods had gone. The farmer-mariners, who daily paddled their flat-bottomed boats to smithy, stores, inn or chapel, had drawn them up for the dry season. The local boys were at their favourite games of rhine jumping, swimming in the calm Brue and hunting peewits' eggs. But the prospect of learning draughtsmanship at Merchant Venturers in Bristol beckoned. It was time to go.

On the last day, wage in pocket, I took the short road home. At the corner I looked back. Thomas Edney was sitting by his workshop door alone. He had picked up a carborundum stone and, white head bent, was meticulously sharpening the points of a fine hand-saw.

Weights and Measures

Introduction

It is not easy to grasp just how difficult it was to practise fair trade when there were a variety of measures and only rough and ready methods of weighing; when sugar arrived in an odd-shaped lump and tea was weighed by the grocer on scales that were worn. As for the farmer, he might use stones that had been roughly chipped to a round weight. There had to be trust and, inevitably, there was abuse. The miller appears to have been a prime offender. Again and again we heard how much more flour was produced from a bag of corn ground at home. But, of course, it was all suspicion; the customer had no method of weighing the corn, and the resulting flour and bran.

Country scales and weights

L. Sanders wrote that a century and more ago country people had to rely on improvisation and the local craftsman for most of their essential equipment, including means to weigh their produce. The Avery Historical Museum has been collecting old weighing instruments from all over the world for a number of years, during which it has acquired many interesting examples made and used in our own countryside. Stone weights are among the simpler of these. Some may be three or four hundred years old, made from stones taken from field or hillside. When farmers had to weigh produce for market and were unable to obtain foundry-made iron weights locally, they sought stones of suitable size, shape and weight and took them to the smith to be fitted with iron lifting rings. Then, by a little chipping or the addition of lead, they were adjusted to compare with a neighbour's weights or with the manorial standards. Hard igneous rocks, such as granite, made serviceable weights reasonably impervious to moisture and capable of withstanding hard wear and exposure.

Occasionally stone weights of the larger denominations, such as twenty-eight and fifty-six pounds, turn up. The large oval one marked '59', illustrated above, would have been used to weigh bales of wool, the extra three pounds being an agreed tare allowance for straps or bindings. This and the twelve-pound weight came from Jersey and were undoubtedly fashioned from large rounded beach pebbles flattened to form a base. A square weight from Shropshire and, though figured '56', weighs only forty-five pounds. This is due not to any dishonesty on the part of the original owner, but to the loss of its lead loading from the large cavity on the under side.

Cart weighbridges, like that at Soham, Cambridgeshire, shown below and platform-scales, an English invention of the mid-eighteenth century, were scarce even in towns and certainly unknown to the farm worker until well into the second half of the nineteenth century. The countryman mostly used beam-scales or hanging steelyards made in the towns by small family concerns employing a few craftsmen and apprentices. Some surviving examples are as crude as those used by the ancient Egyptians four or five thousand years earlier, but others show some appreciation of the fundamentals of the science.

Among the cruder examples are the wooden butter-scales shown below; they are about three hundred years old. A central stand or pillar, turned like a chair-leg on a primitive lathe, carries a wooden beam pivoted on a round iron peg: two wooden bowls or platters are suspended from the ends of the beam. Only the fulcrum pin for the beam is of metal. Scales of this type were used in farmhouses up to the end of the last century.

Larger hanging wooden beam-scales were often part of the equipment of the miller for weighing sacks of grain and flour. They were sometimes as much as six feet long and strongly constructed with metal fittings and rudimentary knife-edges, combining the skills of carpenter and smith. They could be used to weigh several sacks at a time on scale-plates suspended from the end knives by shackles and chains. The wooden beam-scale (overleaf above), is a comparatively small one, about two feet in length, and probably two hundred years old. In contrast, the professional scale-makers of the town constructed their products entirely of metal. Steelyards, based on the principle of the uneven-armed balance used by the Romans and still known by their name, were in common use, for they permitted the weighing of heavy loads without a large number of loose weights. As they required greater precision in manufacture

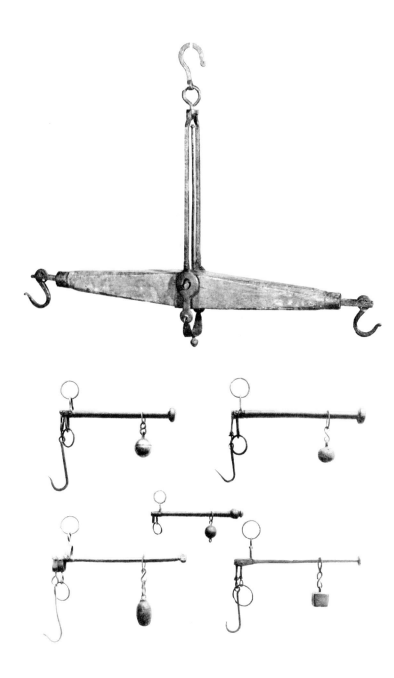

than the beam-scales, few home-made examples survive. The seventeenth and eighteenth-century farmhouse steelyards of English and Continental origin in the Avery collection (opposite below) are small, as steelyards go, and many have wooden arms with metal fittings, poise and knife-edges. Graduation marks are provided by brass pegs driven into the wood at regular intervals. Most of them have two fulcrum knives and duplicate suspensions to take either light or heavy goods—a principle used by the Romans.

An example of a craftsman-made wooden steelyard can be seen in the illustration below of the fine Orkney pundler, which is one of the prized exhibits in the collection. The oak beam is more than six feet in length, and the stone poise weighs thirty-one pounds. Graduation marks correspond to multiples of the Scottish pound. All the metal fittings are of wrought iron, including the knives which are now well rounded by wear. The instrument is believed to have been in use for several hundred years, for the beam bears the weight-stamp of George III obliterating a number of older marks.

Also from Orkney is a wooden bismar or Danish steelyard, used by sliding a cord fulcrum along the counterweighted lever to balance a load. The principle was known to early Aryan tribesmen, who found its simplicity convenient for their nomadic way of life.

The example illustrated above is three feet long and is thought to be of wych-elm. For some two thousand years the bismar, *dhari* or Danish steelyard, as it is variously called, has been widely used throughout the Indian sub-continent and the eastern and northern countries of Europe; but in England it was made illegal in the reign of Edward III in favour of the equal-armed beam and Roman-pattern steelyard.

Cup weights

Readers have been puzzled by objects that have the appearance of weights but are marked with numbers that have no relation to the actual weight. The set of small metal 'cups' belongs to Jack Hill of Leicester. The largest is 2in across and has a hasp and lid, on which are stamped A I (or I) and two symbols, one of which is undoubtedly a fleur-de-lis; the other seems to contain an animal and bird. Inside, on the base, is stamped the figure 16, and others are similarly marked 8, 4, 2 and 1; the two smallest have no figures on them. Their combined weight of 18oz bears no relation to the marking and they came from a long-arm machine.

Wool weights

We handled a 7lb weight owned by Roger Warner; it is of bronze
and still handsome, though worn, with George III's coat of arms
and, on the reverse, the stamp 'V. R.' over a small addition of lead.
Wool weights were originally carried by tax-men who travelled
circuits to weigh and assess wool that the medieval merchants
wished to export. Henry Best, the East Yorkshire yeoman whose
'Farming Boke' was published in 1641, sold his wool on the farm—
it was then carried by packhorse to Leeds, Halifax and Wakefield—
and the merchants visiting farms to buy also carried weights. At
later dates the official weights were found in the hands of the
merchants. J. W. Shilson of Charlbury in Oxfordshire, who used to
buy wool on farms, using the traditional methods of accounting and
weighing. At the turn of the century Mr Shilson joined the firm of
wool staplers with which his family had been associated for at least
four generations; and he recalls his pride when he first went out
alone to buy wool. The advantage of buying on the farm was that
any discrepancy in counting and question as to the price could be
settled there and then, before farmer and merchant parted. As
each fleece was inspected and approved before it was weighed, the
quality or cleanliness could not be the subject of any later claim by
the merchant. Mr Shilson travelled by trap and had with him two
7lb weights and 4, 2 and 1lb weight. His first job on arriving at the
farm was to check that the simple beam scales were in good order
by putting 7lb in either scale—a board 20in by 18in suspended by a
rope from each corner. 'They would turn on a halfpenny.' Sometimes a farmer kept a stone which had been carefully chipped until
it exactly balanced against the two 7lb weights, and this may well be
the origin of the term 'stone' for a weight of 14lb. Others threw any
handy objects on to the scale until they reached this weight. Wool
was bought by the tod—twice 7lb, plus the stone, making 28lb.
The scales held six or seven fleeces and, as it was not desirable to
break one to make an exact weight, short and over weights were
'carried'. So, if five fleeces turned the scale at 1 tod 2lb, a tod of five
fleeces was entered in the book and 'two to the wool' was called, the
2lb weight being placed on the floor near the wool scale. An ac-

cumulation of weights on the wool side was periodically redressed by taking fewer fleeces, weighing less than a tod, and adding to the wool's weights to the wood scale until the beam was brought into balance. If 5lb were required and only 3lb had accumulated, the call was 'two to the weights'. The tax-man travelled on horseback with a pair of weights slung in front of the saddle; but the pair illustrated below, which belonged to Mr Shilson's firm, have a strap, adjustable to an inch or so, that would bring them uncomfortably low on the horse's shoulder. A pair is now a rarity, but single specimens are occasionally seen, and some of the 14lb stones may still lie in dark corners of barns where once wool changed hands. Of the marks the 'A' confirms the weight as avoirdupois, the ewer is the Founders' Company mark, and the dagger is the stamp of the Guildhall. The weights were issued in George III's reign and restamped in 1834.

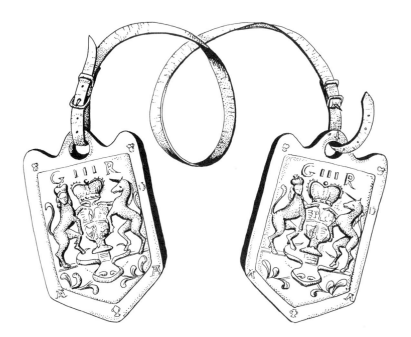

Double-ended measures

The passing of the bushel as a recognised measure of capacity recalled to William Harvey, of Whitstable, the many and varied measures which were used for centuries. 'Well within living memory', he wrote, 'we purchased pears and plums by the peck, apples by the sieve, potatoes by the gallon, nuts by the quart, corn by the strike, and coal and coke by the chaldron (36 bushels). In Whitstable the oyster industry used the tub (21 gallons), the Winchester bushel (21 gallons, $2\frac{1}{2}$ pints), the wash ($5\frac{1}{4}$ gallons), the peck ($2\frac{5}{8}$ gallons), the nipperkin ($1\frac{5}{16}$ gallons) and the bucket ($1\frac{3}{4}$ gallons)'. The photograph depicts double-ended measures reputed to be 300 years old, the two larger of beechwood and the small one of walnut; and William Harvey inquires about their uses. Edward H. Pinto, whose fascinating collection of wooden bygones is now at Birmingham will be known to many readers, has several of the double-ended beech measures, which were used for all kinds of commodities, particularly by grain merchants and fruiterers. In his experience the small wooden cup measures were almost invariably of boxwood; they were used most commonly by apothecaries.

Part Three
DOMESTIC LIFE

Cooking

Introduction

In general the cottager baked his bread and the local cake (dough, lardy and so on) once a week in a specially heated oven, but the rest of his food was cooked on an open fire. In some parts of the country cakes were cooked on a griddle heated in the flame. Communal ovens were a tremendous help, and a village was fortunate if it had a baker who was prepared to open his ovens for a fee.

Regional peasant cooking usually draws its character from the fruits and vegetables that are abundant and cheap in the area. The roast beef and apple pie of England tended to be middle-class dishes. The typical English cottage dish is made from a small quantity of meat chopped finely with potato, turnip, onion or leek and sealed in a pastry of lard or suet.

Bread marker

The marker (right) with the initials M. I., is 2¾in long and made of iron plated with tin. On the upper side is the handle, and on the under side are the inch-high letters, which are ⅜in deep. Mrs E. McKenzie wrote from South Shields, Co Durham, that her grandmother used it to mark the bread and pies she sent to be baked in the community oven. Her recollection is that loaves generally were stamped with initials to ensure their return to the rightful owners. John Higgs had heard of people taking the weekly baking to them, but knew of no specific example other than the two parish ovens at Whatcote, Warwickshire. They were said to have been in constant use sixty or seventy years ago, each user finding her own fuel. On Sundays these ovens were in charge of a caretaker, who heated them and received a penny or twopence for baking a family's dinner. Inez Barratt, who spent the university long vacation of 1929 near Mevagissey in Cornwall, wrote: 'I often climbed the steep hill to look down on the village; and about 9 a.m., if I recall rightly, from one door after another a figure would come carrying a tray covered with a white cloth, and all converged on the village bakery. On Sunday the cloth would cover the joint and on week-days pasties, all of which were marked to ensure that each family got back its own.' Beatrice Martin had vivid memories of appetising smells at Polperro in 1913, as baking tins with meat and potatoes and pudding were carried from the bakehouse. In the previous year, at Horsted Keynes, Sussex, friends of hers bought four cottages which had no separate ovens; the occupants shared a single kitchener in a narrow room at the middle of the row, either using it in turn or pooling resources and cooking together. Gladys Hunkin, of St

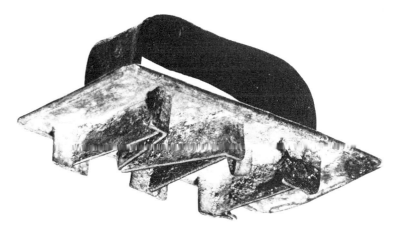

Ives, wrote that the public bakeries widely distributed over Cornwall were not communal, as the users did not provide the fuel or work the ovens. These were kept by men or women who charged fees, and were particularly suited to the cooking of local dishes such as pasties, heavy cake, figgy hoggan and saffron cake. In St Ives the names of at least thirteen of these bakehouses are still known: 'When I first visited the town in the 1920s I saw a woman come from a bakehouse carrying a "star gazy" pie, with fish heads protruding through the pastry. I also recall seeing Mevagissey churchgoers calling for their dinners after service in their Sunday best.'

Hastener

The contraption, 19in long, comes from the kitchen. The lip at the top fitted over a grate bar; the ring is fixed, but the half-circle is adjustable by moving the prop from hole to hole. The pot of food to be cooked or warmed was slipped between the pegs on the ring and the half circle. By moving the prop closer to the pot, the food was tipped towards the fire. In Yorkshire the gadget is called a hastener.

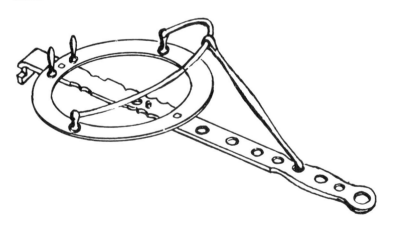

Wafering iron

A pair of iron 'pincers' found in an old house at Lydbury North, Shropshire, were taken to the Clun Town Trust Museum. The

handles are 2ft 3in long, with a link to hold them closed and retain pressure. At the base two facing plates, each 4in in diameter, have the incised designs reproduced above from rubbings. The tool is a wafering-iron used to make thin flat cakes like waffles. The style of the decorations suggests a date towards the end of the 18th century.

Lark spit

The illustration overleaf, from W. T. Jones, shows a lark spit which belonged to Mrs Walter Godwin, of Kencot in Oxfordshire. It was given to her by a former blacksmith in the village. The open frame, which is about 6in high, allowed the birds to be roasted on both sides without being unhooked. In his 'Ornithological Rambles in Sussex', first published in 1849, A. E. Knox describes two methods of catching larks. The professionals used nets with a 1½in mesh, about 25yd long and 16 to 18ft wide and strengthened by lines of stronger cord. At each end was a long and tough pole, held in both hands, and two men advanced with the fully stretched net at an angle of 45° to the ground. On dark nights they would thus sweep the stubbles, clover fields and meadows where larks might be roosting. Another method was favoured by the shooting fraternity. A piece of wood about 18in long, planed off on either side to resemble the roof of a Noah's ark, was mounted with several bits of broken mirror to reflect the sun's rays, and had a central iron spindle.

As it revolved, the flashing light exercised an irresistible attraction to flocks of migrant birds, which would hover within a few yards of the lure, making easy targets. The Wild Birds Protection Act became law in 1880, and in the following year the lark was added to the schedule of protected species. Previously as many as 20,000 to 30,000 used to be sent to the London market at one time. Dr D. A. Bannerman, in 'The Birds of the British Isles', refers to a thousand dozen being taken in a morning on the Sussex downs alone. Country people could then afford to buy little meat and must have been glad of any larks they could get.

Skewer holder

Found on a farm scrapheap in Shropshire, the object illustrated opposite above, was identified by John Higgs as a skewer holder. It came from K. H. Bentley, who tells us that it measures 8in by $7\frac{1}{2}$in and has 'Tradnor' stamped on it.

Salt box

Many readers will remember the salt boxes or jars that were hung
near the kitchen fire to keep the salt dry. The fine oak specimen,
below, is 11in high, with a leather hinge that will not corrode.

Pot quern

The object, shown much reduced above, was found in Litton Dale and removed to Burnsall in Wharfedale, where it was photographed by G. Crowther. It now serves as a garden ornament on the side of the road through the village. Dr E. Cecil Curwen wrote: 'This is certainly a pot-quern. I take it that the upper stone is in position and could be made to revolve inside the lower one, though the usual handle is not shown. At the apex is a kind of hopper by which the material to be ground can be fed into the quern while the upper stone is being rotated. The meal is removed from the lower stone through the opening seen on the left in the illustration. I have little evidence as to the distribution and range of date of pot-querns in general; my own interest has been mainly in pre-Roman querns. I suspect that they may be of German origin, and believe they first appeared in Britain during the Roman period, possibly introduced by German auxiliaries. There is some evidence also that they were used here by the Saxons. For the medieval period evidence is scanty, but they were almost certainly popular then. Many of the pot-querns found in Britain are made of Niedermendig lava from the Rhineland. It has been suggested that pot-querns were not necessarily used for grinding only corn but also for pulverising other things, including vegetables.' In the 1920s M. J. Phelps saw in the Gower another form of grinder. A round stone, about the size of a cricket ball stood against the door as a stop. An old lady told him they had ground mustard and sometimes pepper with it. In her youth mustard was always ground from seed on a flat, slightly hollowed stone with any handy-sized pebble with one good surface. The pebbles were often used as found but were sometimes worked a little to make grinding easier.

Flour boulter

The flour boulter illustrated on this page came from Norfolk and is of a type that was familiar in farmhouse kitchens a century ago. The meal was emptied at the top on to a sloping platform and filtered on to a brush set at an inclined plane. The flour fell into the drawer at the bottom, and the middlings came out of the shute at the side. The two top drawer-fronts, one of which is without handles, are dummies. Allan Jobson, who sent the photograph, wrote: 'In a family of 14, half a coomb [2 bushels] of wheat would be sent weekly to the mill to be ground, and it was alleged that, if the resulting meal was refined not at the mill but at home in one of these boulters, a saving of one shilling and 4lb of flour was effected'.

Food warmer and pap boat

The picture of the glazed pottery 'mug' with side opening reached us by way of the Ulster Folk Museum, Belfast. Within it is a circle 2½in across with raised rim, and on each side of both handles is a set of four holes in diamond pattern 4in from the base. On the bottom of the 'mug' is the figure 20. It was certainly intended to take a small night-light and probably used to warm food. This brought a letter from Brittany. 'Here such things,' Frances Richardson wrote, 'are still used in the older houses as a kind of samovar with a squat and bulbous teapot sitting on top. Generally the pot infuses tisanes made of *tilleul* (lime blossom), cherry stalks, eucalyptus leaves or any of the herbs which are bought from chemists and have to be infused for about ten minutes over heat. In the bottom of the warmer there is a small flat china saucer, in which a night-light can be placed; but more often a table-

spoonful of thick oil is floated on the same quantity of water. Wicks threaded through half-inch squares of tin are sold by the handful in hardware shops; they give tiny flames—just enough to keep the pots warm'. Edward and Eva Pinto told us that the food warmer originated in France, where it was known as a *veilleuse*; they have examples in wood, lined with brass or iron, in their collection. A doctor who trained in Belfast recalled that food warmers were in common use there eighty years ago; a shallow dish with two lugs rested on top and was mostly used for milk. Mrs Richardson too, as a child in England, saw babies' milk warmed over a night-light in a thick earthenware container.

Below are a pewter pap boat about 6in long, used for feeding slops to children and invalids, and a baby's feeding bottle with a blue transfer print of a country scene; this was given to the present owner when she was first in service more than seventy years ago.

Scoops

Below are two apple scoops, each measuring about 5in. One, with initials and date (1806), is made of box-wood and came from William Morris's village of Kelmscott. The other is a sheep-shank fashioned with a penknife by the uncle of George Swinford, of Filkins, to whose mother it belonged. He recalled that before the days of dentures these scoops were much used by people who had lost their teeth. His mother used to scoop out an apple, leaving the skin intact, until it would crumple in the hand like paper. The illustrations brought a letter from Professor Raymond Dart of the University of Witwatersrand, Johannesburg, who was working on tools used by man-ape (*Australopithecus prometheus*) in the Transvaal. Similar scoops made from the cannon bones of antelopes were in use there nearly a million years ago—not for apples, of course, but for soft meat, liver and brains and so on. Professor Dart has also found similar scoops at Makapansgat in the Transvaal, where they were used by middle stone age man about 15,000 years ago. Lilian Hayward recalled a collection of five sheep-bone scoops in Shropshire. Four were found in homes at Clun, and the fifth was bought from a Birmingham dealer who said that it had come from the same place. They are not unlike the bone scoop at Filkins, except that the opposite ends from the scoops are much indented and were apparently used to peel the fruit. At Clun they

are called 'oloppers'—a strange name of which Mrs Hayward failed to discover either the significance or the derivation.

E. A. C. Husbands, with sharp eye, rescued the marrow-scoops (below) from the big mustard-pots in a workmen's cafe. He spotted another in an inn where they serve a farmers' ordinary. Silver scoops of the eighteenth and nineteenth centuries are now popular collectors' pieces, but those made of bone are less often seen or, perhaps more accurately, recognised. About 9in long, they were used at table for extracting marrow from bones; the channelled handles fitted the smaller cavities. An early edition of Mrs Beeton's 'Household Management' includes this recipe: 'Have the bones neatly sawed into convenient sizes and cover the ends with a small piece of common crust, made with flour and water. Over this tie a floured cloth, and place them upright in a saucepan of boiling water, taking care there is sufficient water to cover the bones. Boil them for 2 hours, remove the cloth and paste, and serve them upright on a napkin with dry toast'.

Mould

Gingerbread cakes were often baked in farmhouses and cottages, gilded and sold at fairs; hence the name 'fairings'. Many of the moulds used in their making had attractive traditional designs, such as the two birds on this beechwood mould, 8½in by 5in, above overleaf, from East Hendred.

Cutters

Below overleaf is a pair of sugar cutters whose turned wooden

handle and base (10in from end to end), and some elaborate design in addition, suggest that they were used at table rather than in the kitchen. White sugar used to be made in a cone about 36in high and 14in across the base. The householder bought large pieces broken from this and used cutters to reduce them to manageable size.

Nutcrackers

In the attic of a Finstock farmhouse in west Oxfordshire, where Dorothy Bolton's family lived for 200 years, she found the 6in nutcracker (above) decorated with a band of carving illustrated above. In the last century, when nuts were more popular, it was as customary to carry a nutcracker as a penknife, though this specimen seems large for the pocket. The hazel-nut crackers below were given forty years ago to Kay Shimmer by J. Tuttle, an old Sussex hurdle maker, who used to carry a pair when working in the woods in autumn. After shaping a straight piece of wood with his knife, he soaked it well, then doubled it over and bound it tightly with a thin strip of split hazel until it had dried out.

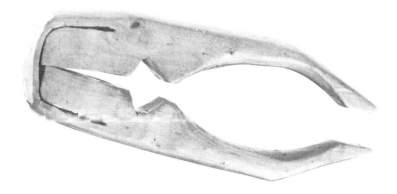

Swinging churn

The Museum of English Rural Life now houses two hundred
folk items from the Sharp collection at Wonnersh. But—a warning

to us all—the Rev C. J. Sharp failed to make full notes on use and origin, and the 9in wooden container, illustrated left was one of the puzzles. It has weep holes in the base and was found among dairy equipment. It suggested a primitive form of centrifuge to Peter Rosser of Hurstbourne Tarrant. Edward and Eva Pinto were able to confirm this: 'It is a swinging churn, probably the earliest type of churn in Europe. An almost identical one in the Pinto Collection, now at the Birmingham Museum, was excavated from an Irish peat bog. Both are almost certainly medieval.'

Butter by the yard

When talking of her childhood Sydney Hore's mother often recalled buying butter by length, apparently in long thin rolls. This would have been in the Potton district of Bedfordshire in the 1860s. Was the method, he wondered, used only at some small farm without scales ? Miss Enid Porter, curator of the Cambridge and County Folk Museum, sent the photograph (overleaf) of Mr Smith selling butter at a Cambridge door in the early years of the century. He supplied butter to dairies and private customers. Miss Porter wrote: 'It is generally assumed that the practice of making up butter in yard lengths, of approximately 1in diameter, was for the convenience of college butteries, where the undergraduates' sizings could thus be gauged more quickly. After the Peasants' Revolt of 1381 the King punished the town of Cambridge for its part in the riots by removing certain of its privileges, among them the testing of weights and measures, which until the Cambridge Award Act of 1856 remained in the hands of the University. There is still carried in University processions an object known as the yard-butter measure, though its shape suggests that it was more suitable for estimating the depth of liquids in casks. Yard butter disappeared with the coming of cheaper butter from abroad, though my grandmother continued to buy it from a farm till the early 1920s'. L. G. Jacob wrote of undergraduates in Christ's and other colleges being supplied from the buttery daily with a loaf of bread, 3in of butter and a pint of milk. 'The arrangement', he added, 'was said to date from earlier times when some students, eager for

mental nourishment, cared not if their money was inadequate for bodily nourishment; the University authorities ensured that they would not completely starve'. Several correspondents recall purchases of butter in this form in Cambridge market at the turn of the century. P. Fordham described it as 'pale as a primrose and unsalted'; the only price he can remember is 2s a foot. Mrs D. M. Vinnicombe recalled how about 1900 she was sent to the shop in the little village of Aldreth, near Haddenham, to buy half a yard of butter and saw there 'quite a pile of rolls, each about 1in across'.

Curd mould

Miss G. M. King, Rural Home Economics Instructress for Oxfordshire, asked how the china shape, above was used. It measures 7½in across. The perforated mould was recognised by several readers as one used for draining and shaping curd dishes. In 'The Cook and Housewife's Manual' (1826) Mistress Margaret Dods gives this recipe for Hatted Kit: 'Into the kit [originally a wooden milking-pail] put two quarts of fresh good butter-milk and a pint of milk hot from the cow. Mix well by jumbling; and next milking add another pint of milk. Remove what of the hat [curd] is necessary.

This dish, if to present at table, may be moulded for an hour in a perforated mould and strewed over with a little pounded sugar and then nutmeg or cinnamon'. Mamie Molloy told us that in France a similar dish, made with sour cream and drained in a heart-shaped mould, was known as *coeur à la crème*. Betty Johns sent a modern version of the dish, in which the milk is curdled and flavoured with fruit: '*A Curd Star*—a charming old-fashioned English dish, for which a star-shaped mould with holes in it for draining is needed. These are apparently no longer made, but it is the work of a moment to pierce a few holes in a cheap star-shaped cake-tin. Put a slice of lemon peel and a pinch of salt into two pints of fresh milk and bring to the boil. Add two tablespoons of caster sugar, a sherry glass of sweet white wine, Marsala or sherry, and four whole well-beaten eggs. Boil until the eggs curdle, then simmer gently until you see the solid part separating from the whey. Leave to cool a little, and turn into the mould to drain, removing the lemon peel. Leave until next day, when the curd should be quite solid. Turn out on to a plate and serve with fresh thin cream poured over it'.

Home Industry

Introduction

A hundred years ago there was little furniture in the overcrowded cottage. The cottager made do with a bed bolted together, while a curtain across one corner of the room hid his clothes. Downstairs there was a table, the Windsor chair for father and kitchen chairs or stools for the rest of the family. Against the wall stood the chest of drawers or cupboard which held all the family impedimenta. A place also had to be found for the tools of the industries carried on in the home. Straw plaiting, smocking, button- and cord-making, knitting and gloving were all important cottage industries that have left their trace.

The machine of the town manufacturer killed the cottage industries and, until the works' bus arrived in the village, the cottager was again dependent on the state of agriculture.

Straw splitters

The tools illustrated overleaf came from a village near Leighton Buzzard and have been in possession of Edward Simpson for many years. He wrote: 'In the early years of the century, when straw hats

were recognised summer wear for men as well as women, the town of
Luton was a centre of manufacture, and there was a great demand
for selected straws plaited in various degrees of fineness. A man with
a pony and cart used to deliver bundles of straws to a skilled plaiter
in each village. She in turn distributed them to the women for
plaiting in their homes and collected the finished work for the man
to pick up on his next visit. The women used these tools for splitting
the straws evenly to the required number of strands for plaiting in
the various grades. The sketch shows one with five blades and
another with four, which is also shown in a partly split straw. The
tools are made from what appear to be simple brass castings with
the cutting blades filed to a reasonable degree of sharpness; the
edges need not be very sharp to split a straw satisfactorily. The spike
acts as a guide when pushed into the end of the straw'. The head was
sometimes fashioned from bone. The splitter at the bottom has an
oak shaft $3\frac{5}{8}$in long morticed into the bone. It came from an old
countryman, Mr Groom of Acton near Sudbury, whose grandmother used to 'earn a few coppers by it'. She was born in 1823, and
the tool belonged to her mother before her.

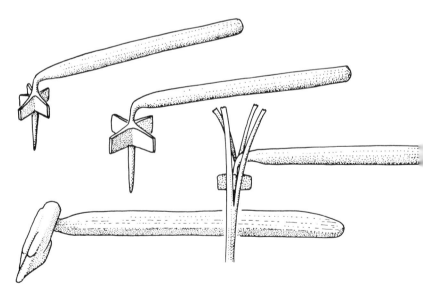

Clamps

The gloving donkey was a form of vice that held the material
while the glover sewed along the seams. After stitching the glove
was released by depressing the foot-pedal. The example illustrated
below left is 34in high and came from Moreton-in-Marsh, Gloucestershire. A much smaller table clamp (below right), the overall
length is 3½in came from Suffolk. It is made of brass; the small
piece in the angle of the 'C' is pivoted. While the seamstress pulled
the material away from the clamp the pivoted arm would grip it
tightly, but as soon as the tension was eased, the arm released its
hold and a fresh length of seam could be slipped through.

Knitting sheaths

The two 8in wooden knitting-needle holders were picked up at a local sale by Mrs Cordukes of Whemby, York. The slot fitted over the knitter's belt, and the needle was fixed in a hole at the end of the sheath, so that the wearer had a free hand to manipulate the wool. These sheaths were often given as tokens of affection, elaborately carved. The two balls in that dated 1818 move freely in their open-sided box; they were carved out of the single piece of wood and are purely decorative. On the north-east coast many sheaths

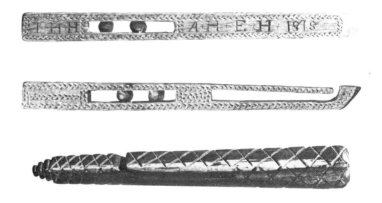

were made by seamen who based their designs on objects seen on their travels. The thick jerseys worn by fishermen were heavy and in those parts the sheath was often held under the arm to ease the weight of the knitting. In Cumberland the sheath was common, but Mrs J. Musgrave of Windermere remembered her grandmother knitting with a needle held in a pad of oat straw, about the width of two fingers and 6in long, that was tucked into her apron or skirt waistband.

Pegotty

The circle of wood (opposite), an inch thick, with thirty-two 'nails' protruding about 1½in from it, measures 9in across. Mrs F. M. Longman, of Wombleton in Yorkshire, obtained the loan of

it from an old lady who lived for a long time on a farm in the North Riding; this and a smaller one with twenty-four wires came from her mother's home, but she does not remember having seen it in use. A pegotty or a knitting Nancy usually have wooden pegs. The method of use is much the same as that adopted by children who stick pins in the tops of cotton-reels and make long tubular woollen strips with them. Pegotties came in different sizes, presumably according to the measurements of the garment required. Forty years ago they were quite common.

Lucet

In 'The History of Needlework Tools' Sylvia Groves gives clear instructions for working a square cord on a lucet. The embroiderer's workbox would have contained an ivory, pearl or tortoiseshell lucet; the 18th-century cottage tool was less ornate and usually of

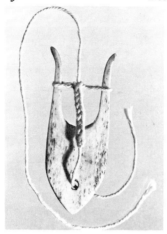

horn or wood. This lucet (above), belonging to William A. Cocks of Ryton-on-Tyne, is of bone and $3\frac{1}{2}$in long. The firm knot produced on a lucet was widely used for making non-stretch stay and shoe laces until machine-made cord was introduced about 1850.

Staymaker

Joan Cope wrote that the 1880s in a remote dale near Wastwater the dressmaker specialised in making corsets. 'She would walk to the isolated farms and cottages to measure and fit her customers, and thought nothing of a ten-mile journey over the fell tops from one dale to the next. The corsets, at that time known as stays, were made of a drab-coloured material called cantoon, or jean, and were sewn on a machine with casing cotton, similar to what we now call linen thread. If the village shop ran out of whalebone, she used instead slivers of oak from the basket maker. The front piece of stiffening was three inches wide, and the top perforated to take a binding of linen tape; the edge was very hard. The local shoemaker then put in eyelets for the laces. The stiffness of these stays when new was quite daunting: sitting down entailed consideration, manoeuvre and effort, and bending was accompanied by much creaking and groaning of the supports. How women so encased managed to do the housework is difficult to imagine.

This staymaker received 1s 6d a day and her meals when she visited houses, and charged 7s 6d for the stays irrespective of size. One customer had a forty-inch waist and every pair of stays meant a financial loss to their maker, who dared not offend by a refusal to supply. A neighbour who is now seventy-eight used, as a little girl, to carry the stays to the shoemaker, wait while the eyelets were put in and then take them back, for which service she was paid a penny. No doubt more amenable garments could have been bought in the nearest market-town, but that was more than fifteen miles away, with horse and cart the only means of transport.'

'Eggs'

China or pot eggs, those deceivers of hens, will soon be part of history; hens, alas, now have little opportunity to exercise their bird brains. But there are other egg-shaped objects, with quite different uses, that also belong to the past. Richard Brandon of Buerton, Cheshire, owns a boxwood egg, almost $2\frac{1}{2}$in long, known to be a hundred years old. This light but dense material was easy to hold when sewing; it is a darning egg. Eggs made of heavy smooth marble were used as hand-coolers by needlewomen working on fine white work and by nervous young ladies anxious to give an impression of sangfroid. E. A. C. Husbands came across another egg, when he was sheltering from a sudden and severe storm in the wilds of Derbyshire. He noticed 'a polished egg-shaped piece of granite, $2\frac{3}{4}$in long, standing on a mantlepiece'. The elderly women of the house told him that it was a comfort stone, and that it was given to a woman in labour to grip in her right hand.

Dorset buttons

In the eighteenth century the cottage industry of 'buttony' was widespread in north-east and east Dorset wrote Marian E. Chappell. About 1690 Abraham Case, a native of Shaftesbury, invented a new type of button, possibly to circumvent a law of 1685. In order to protect the metal-button trade in Birmingham, this forbade the making of cloth-covered buttons; and it remained on the statute book until 1727. The new type of button, known as 'cloth-work', could scarcely have been called 'cloth-covered' because, although it had a linen centre, it was covered with fine lace-stitchery. Refugee Huguenot lace-workers in the north-east of the county took to the work, and their fine lace-thread was used in many of the early buttons.

The first cloth-work button, based on a ring of Dorset sheep-horn, was known as a 'high top' (below left). A piece of linen cloth was pulled in a twist, so that it formed a firm conical shape with the horn as its base. Over the linen was worked a kind of chain stitch which went round and round, the loops being so arranged that they came underneath each other down the sides of the button from tip to base. The secret of making this button has been lost, and indeed how the smallest were contrived at all is a puzzle. They ranged from about half an inch high and a quarter of an inch across the base down to something like one-fifth of an inch high and three-twentieths across. The high top was popular for gentlemen's hunting waistcoats and would last a lifetime.

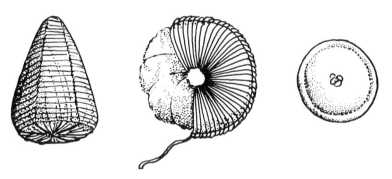

Next, about 1715, came the 'Dorset knob', a development of the high top; although much flatter and broader, it was made in much the same way. The buttonhole ring or 'bird's-eye' (above centre) followed. This was quite stiff and firm but, when taken apart, has been found to be based only on a piece of cloth cut on the bias, firmly twisted, formed into a ring and tightly buttonholed over. It may perhaps be regarded as a transitional type of button, the ring of horn having disappeared.

Peter Case, a grandson of Abraham, invented a quite different type of button about 1750; it was based on a wire ring made of a special alloy which did not rust. First came the 'singleton' (previous page right), slightly padded in the centre and covered with linen cloth. Some were back-stitched round on the inner side of the ring, some had two rows of back-stitching and others, more rarely, were buttonholed round over the ring. In the middle were different patterns composed of groups of french knots, which might be from three to seven in number.

The singleton, probably used largely for shirts, led directly to the button which became so widely known for a century and is still recognised as the typical Dorset button. It is based on a wire ring and has a worked centre. The wire ring was made in many cottage homes in the north and east of the county, and even over the border in Somerset and Wiltshire, during the second half of the eighteenth century and the first half of the nineteenth. The wire was brought from Birmingham in great wagons with broad-rimmed wheels which could carry loads of a ton to thirty hundred-weight. It was cut into various lengths and then made into rings by specially trained boys and girls known as 'winders and dippers'. They twisted the wire round a spindle, then soldered the ends together. The rings were tied into gross lots by 'stringers'.

The ring was covered with buttonhole stitch (see over); this was called 'casting'. The knots on the outside were then smoothed inwards by children with a wood or bone 'slicker'. Thread was wound across the diagonal of the ring again and again, care being taken to cross exactly in the middle. This was known as 'laying'. Then the crossed threads were caught with two stitches in the middle, one up and down, one across, to fix them. Finally the centre was filled in with one of several designs, which gave the buttons their names. Some were called 'Dorset crosswheel', 'old Dorset', 'Blandford cartwheel', 'honeycomb' and 'basketweave'. There was a name for each size too. The tiniest, a bare three-sixteenths of an inch across, was a 'mite'; some of the larger ones were 'waist-coats' and 'outsizes'. Buttons which became dirty in the working were boiled in a linen bag. The finished products were mounted in gross lots on cards covered with paper of different colours according to quality. Yellow was for the cheapest, dark blue for a better quality and pink for export only. Very few of the pink remain today.

Between 1830 and 1850 the button makers were paid 1s 8d to 3s 6d a gross according to quality, the finished products being taken by women to certain picking-up spots or to agents. In 1952 an old lady then ninety-two, told how women used to walk ten miles from Margaret Marsh to Blandford with buttons; half way there was a house where they called for a rest and bread and cheese and beer usually paid for by the agent.

There were agents at Shaftesbury, Blandford, Sherborne, Bere Regis, Poole, Langton Matravers, Tarrant Keyneston and elsewhere. At the Milborne Stileham agency, set up by Peter Case Jr in 1803, buttons were accepted each Friday, when the place is said to have been crowded 'like a fair'. In the more remote districts agents called at fixed times at collecting points. Payment was not always made in money, but sometimes in goods.

Towards the end of the eighteenth century Lady Caroline Damer of Milton Abbey established a school for twelve poor children, who were clothed and taught reading, spinning and buttony. In 1812 a Mr Acheson is said to have been the chief employer of labour, to the extent of 1200 women and children. During the first three or four weeks children received no pay, for they 'spoiled much thread'. They then received a penny a day for two months, and a shilling a week for two more months. Eventually the best hands could earn as much as 10s to 12s a week. Buttony was unpopular with farmers because it was difficult to get women and children to work for ninepence a day in the fields.

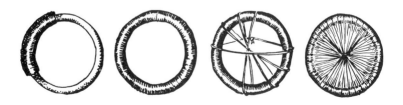

In the 1840s a London office in Addle Street had an annual turnover of £10,000 to £12,000. In addition to sales in England, Scotland and Wales, buttons were exported to all the principal cities of Europe and to Boston, Quebec and New York. A member of the Case family who had gone to Liverpool to oversee the export trade made a fortune, which enabled him to build Case Street and Clayton Square in that city. Mrs Jackson of Shaftesbury, who has done so much in the past fifteen years to revive the craft, received from America a sketch of a card which appeared to show the coasts of England and America with ships on the ocean between them. To different parts of the card, on the ships, sails and so on, various samples of Dorset buttons were fastened. She believes that an agent sent this out as a sample card.

With the invention, by one Ashton, of a button-making machine which was shown at the Great Exhibition of 1851, trade fell away dramatically. The last large order fulfilled by a Case is said to have been one valued at £850 in 1859. By the end of the century the cottage industry had almost completely disappeared. All memories of it, too, might have vanished, if Florence, Dowager Lady Lees had not then sought to revive the craft at Lytchett Minster. While people who had been actively engaged in buttony were still alive, she scoured the country for information and had all the traditional wire-rings designs copied. Variations were also added. The 'spangle' was a cartwheel with tiny sequins round the rim; the 'gem' resembled a wooden bead wrapped round with silk thread, then embroidered longitudinally. The 'yarrell', a development of the crosswheel and some inch and a quarter across, was used as an election button in red or blue. The bird's-eye of Edwardian times was formed on a shaped circle of wood with a hole at the centre, instead of a ring of twisted cloth. Many of these Edwardian buttons were made with silk thread, and orders were often given for the materials for covering dresses. The trade in them was good until the 1914–18 war. I have a vague recollection of seeing cards of buttons for sale in a tiny cottage window at Lytchett about 1910.

In 1931 Mrs Jackson and her husband bought for conversion

two cottages in the hamlet of Twyford, near Shaftesbury. This small place has, as she puts it, neither church nor shop, post office nor public house; yet her cottages were numbered 54 and 55. This high numbering made her curious, and inquiries among the inhabitants disclosed that, a century earlier, this had been a large and thriving village devoted to buttony. The collapse of the craft had brought poverty and starvation to Twyford, where whole families had been engaged in it. So acute did the problem become that the Government transferred many of them—350 people from the Shaftesbury area alone—to Australia, Canada, Tasmania and other oversea territories. It would be interesting to know of any traditional stories told by descendants of these families. Their deserted English homes eventually crumbled away; but in various places, before war-time ploughing, garden trees and shrubs marked the sites.

Ever since, Mrs Jackson has been acquiring information on Dorset buttons. Four years ago interest in the subject was given a further impetus when an elderly lady who was giving up her home handed over a trunk full of the buttons, antique and modern, to Madeline Lady Lees, the late dowager. She had had this collection ever since the earlier Lady Lees had bought up the remaining stock when the last agency came to an end in 1908 on the death of old William Case, the last surviving member of the family that had invented and promoted the Dorset button.

Now it is possible to buy both antique and Edwardian buttons; and not only these. Mrs Jackson and her pupils make modern buttons to match a wardrobe; and you can learn to make them for yourself.

The Weekly Wash

Introduction

The widow who took in washing is a well-established figure of literature, and this was indeed one of the few ways by which a woman could make a living for herself and family. The woman with flair might expand the business and employ others. We received details of the routine and discipline of how the work was done: Monday collection, Friday deliveries, buildings full of steam and dripping with water and happy, laughing sessions when the laundresses curled their hair with the tongs that were bought to get up the starched frills.

Many of the aids that have puzzled readers belonged to the professional washerwoman, whether she worked at home, ran a family business or was employed in the laundry of a great house. The copper was the great labour-saving item for the cottager's wife. The old copper was a stout brick affair; the firebox below the ample basin, led into a well-built chimney. It was often housed in its own small shed and sometimes a simple wash-house would serve a group of housewives. Each had her allotted day and, in bad weather, she could hang the wet linen round the copper fire.

Washing machine

The washing machine (below left), made probably between 1860 and 1870, operates on the same principle as a butter-churn. G. C. Farmer found it in a farmyard on the outskirts of Keswick. The stand, rather like that of a modern wringer, supports a three-gallon copper container pointed at each end. This was opened in the middle for the insertion of the washing, then closed and tightened with wing-nuts, before being turned by a handle, again rather like that of a wringer. Each time the container travelled through half a circle the clothes dropped some 2ft to the bottom. It was up-ended for emptying by means of a tap which appears on top in the picture. A small plate gives the name of the makers: The Torpedo Washer Company, Huddersfield.

Lye dropper

Before washing soda came into use, lye made from white-wood ash was used to soften water for washing clothes. A lye dropper (above right) was balanced on a forked stick of hazel or maple on top of the tub, and twigs were arranged in the bottom of the dropper and a clean cloth spread over them. Ash from the brick oven, copper or open fire was then placed on it. Water was poured over the cloth and as this dripped through it took the alkaline salts with it. The resulting lye was strained through muslin to remove any ash. The dropper illustrated, 20in square and 9in deep, came from a house in East Hendred, Berkshire, where for three-quarters of a century it had remained unused.

Mangle bat

Mangle bats were very common in Scandinavia, where they were beautifully made either gaily painted or decorated with carvings. Often the handle had the shape of a horse. The only specimens John Higgs had found in English museums have proved to be of Danish origin, but it would be surprising if they had not at one time been in common use in the eastern counties of England. Mrs Dudman of Whitby wrote that, before her mother could afford a mangle, she used what she called a battledore, which is evidently another name for a mangle bat. 'It was a large oak piece 3 or 6in thick shaped like a cricket bat but much larger. All the plain clothes which needed ironing were wrapped round a roller, and with the very heavy battledore they were rolled till the creases had gone.' It appears then that English battledores were not decorated like the Scandinavian ones, and this may explain why they have not survived as museum pieces. The word battledore is interesting also, for it is usually associated with the game, but according to the Oxford English Dictionary its first use was for a 'beetle used in washing, also for mangling linen clothes'. The Danish mangle bat illustrated overleaf dates from the early eighteenth century (1735) and has a very beautifully carved top and handle. It is 25in long and the horse $4\frac{1}{2}$in high.

We learnt that the mangle bat was in use by the Royal Navy as late as the 1920s. 'When a ship was newly commissioned', wrote E. W. Taylor of Didcot, 'each mess was issued with a set of mess traps, but this did not include bat or roller. So the senior hand would scrounge a piece of hardwood, preferably oak or teak, from

which he would carve the bat with his pocket-knife or "pusser's dirk". It was shaped like a cricket bat, about 3ft long, 3in to 4in wide and 1in thick, and was sanded and scrubbed to ivory smoothness. The roller was a 2ft section cut from the loom of a broken oar, again sanded and scrubbed; it was also used as a rolling pin for pastry in days when all food was prepared by members of the mess and only cooked by the ship's cook.' H. L. B. Bolton of Deddington in Oxfordshire also used a 'bat and fid' during his years in the Royal Navy, fourteen of which he spent on the lower deck; there the bat was used weekly by each man in turn. 'Duck suits were wrapped round the roller as taut as possible and laid on the spotlessly scrubbed mess table. Then, one hand grasping the handle of the bat and the other pressing hard about midway, the bat was given a sharp push and the roller was rolled under pressure between bat and table. Some half-dozen rolls sufficed for ordinary wear, but on a station where tropical kit was worn the bluejacket's "tiddly" or shore-going whites came in for special treatment. They were made of drill and usually specially cut for him, with edging and collar of blue jean. The suit was turned inside out, and leaves torn from a glossy magazine were placed carefully in direct contact with the whole of the outside of the suit. "Jumper" and trousers were then folded, and each piece was rolled as much as a dozen times. The gloss was transferred from paper to suit, which had a finish almost up to the standard of linen from a Chinese laundry.' A. C. Cochrane of Dorking sends a reference to a 'battling stone', which was a large smooth stone set in a sloping position by the side of a stream. In the North washer-women used to beat their linen on it.

Crimping

The goffering stack (above left), from Buckinghamshire, stands 13in high and was used for crimping lace ruffles and trimmings for caps and bonnets. The damp material was wound round the loose wooden quills, clamped in place and put near the fire to dry.

Zillah Halls of the Nottingham Museum and Art Gallery identified the more elaborate crimping-machine (above right), which William Cocks bought at a sale in Northumberland. Near the top of the wooden frame, which is 20in high, is a roller with hooks to hold the material to be pressed. The inverted T-piece runs in a groove on either side and is fixed in position by a wedge in the top bar. Missing is a bundle of sticks, which were placed horizontally in the space between the T-piece and the base, their ends fitting into the grooves in the two uprights. They were of such thickness that, when fitted, they formed two tiers, front and back

As they were inserted, one after another, the slightly damp material was twined round them to give the desired corrugation. Finally the T-piece was pressed down on to the sticks, and after a day or so the material would be both dry and crimped. Miss Halls had not seen a roller with pulleys, presumably a refinement for holding any surplus material out of the way.

The crimping board was harder to use but gave a sharper finish. Edward Pinto writes in *Treen* (Bell, 1969): 'Crimping boards and their correspondingly grooved or serrated rollers were used for forming the minute crimpings, gathers or ruckings on 17th- and early 18th-century fabrics. The fabric was first starched, then damped, placed on the board and rolled; when used for gathering or rucking, the pleats were subsequently stitched at the top. They are usually of boxwood.' Kathleen Stevenson of Broughton Beck, Ulverston, sent us the photograph below of a board and 8in rolling pin that acquired a secondary use. She believes that gunpowder for the Coniston copper mines was ground on the ridged board; metal was ruled out because of the danger of a spark. She has been told that the gunpowder was funnelled into a hollow straw, which was also the fuse. Perhaps the Coniston men took the board to the mines when it was no longer required in the house. The polish could have been the result of hard wear in the new use.

Irons

Raymond Cripps of Witney, Oxfordshire, sent the photograph
below of a charcoal iron which is of a type common during the late
19th century. To get up heat the ventilator at the back was
opened and the iron swung in the hand; the funnel carried away
the fumes. A very similar iron was still in use in the second half of
the twentieth century, in the Abruzzi in central Italy.

Ten-year-old Kevin Grant of Wootton, near Woodstock, found the iron illustrated below under some old bales in a spinney. 'We were amazed at the sight of it', he wrote. 'It had strange holes in the sides and a hole at the back. I thought it was an old-fashioned steam iron.' He was almost right, for it was identified at Reading as a spirit iron, 'about as safe as a bomb'.

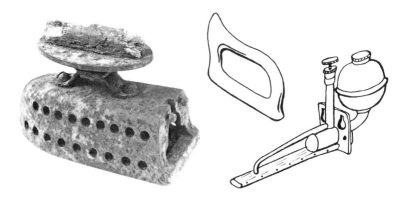

The little laundry maids

Amoret and Christopher Scott wrote: when we acquired our sixteenth-century half-timbered cottage in Worcestershire, our predecessors showed us the foundations of a large building in the paddock behind the house, and a small leather-covered book which had been passed on to them by the previous owners. The site had been occupied by a laundry which was run as a training school for young girls from local institutions. The book had been used by the matron of the establishment, who lived with her staff in our cottage, to record the advent, progress and subsequent fate of each of her charges. Their average age on arrival was about twelve, though in 1888 one, Edith Crane, was admitted at the age of seven. All had originally been taken into institutions with a background of broken homes: dead, crippled, insane or worthless parents. A common formula is 'Father dead, mother deserted her'.

From 1870 until 1925, when the home was finally closed, there is a complete record of the running of the laundry. The girls washed all the linen of the great house which patronised it, and neighbouring houses could have their washing done at reasonable prices. In 1905 damask table-cloths up to three yards long were laundered for 6d; nightdresses cost 8d, 9d or 1s 6d depending on whether they were white, flannel or silk.

There seems no doubt that the establishment was well and efficiently run according to the somewhat stringent principles of the time and circumstances. It was inspected annually, and there were regular vists by the chairmen of the unions from which the girls came. While the reports inscribed in the book are uniformly complimentary, one cannot avoid the picture of the pathetic twelve-year-olds who lived and worked there, pummelling the huge sheets and tablecloths in an atmosphere of steam for most of the hours of daylight (and a good many dark ones in winter) and packed off to bed in the long bare dormitories. In some of their reports the inspectors do remark that the girls seemed to be working over-long hours. Their pleasures were obviously few. Country dancing was allowed once a week; those who could sing at all attended choir practice, and all were scrubbed for church, where attendance was compulsory three times on Sunday.

If the girls had a thin time of it, the matron's patience must often have been sorely tried. 'Five girls were so rude to Mrs Kitching that they were punished by having no pocket-money and no eggs for breakfast. They object to doing their work over again when not properly done.' 'Florence Ratcliffe's character not very satisfactory. She cannot get up in the morning.' 'Mary has to remain out of the laundry, she is so naughty.' One particularly unfortunate affair involved Trumper, the boiler man and only male in the establishment, Mrs Kitching the laundry matron and Ethel Anderson, one of the girls:

'25th July, 1893: Trumper has been very troublesome this week in putting the girls up to mischief and sauciness. He is also a great deal too familiar with them, especially Ethel Anderson.

'28th July: Matron has had a serious talk with Ethel. She had, however, been very impudent to Mrs Kitching and was sent to bed in consequence.

'4th August: Trumper has been given notice to leave, chiefly on account of the girls.'

It must be added, however, that the rebellious Ethel went on to make good and was finally placed as housemaid in an unknown mansion, whence favourable reports were received.

Not all of the girls survived the rigours of the laundry, and even the first laundry matron, in 1871, became 'mentally affected'; eventually she drowned herself. Several girls are recorded as being untrainable and were returned to the institutions from which they came. One complete page of the book contains a name, date of arrival and the laconic statement, 'Ran away'.

When they had made their own 'trousseaux', an essential part of which were two pairs of black knitted stockings, most of the girls were placed in the great houses of England at wages ranging, in the 1870s, from £12 a year for a fifth laundry maid to £14 for a second, plus 2s 6d a week beer money. Those not sturdy enough to do laundry work were found other forms of domestic employment at reduced rates: in 1874 Caroline Hall was sent out as an under-kitchenmaid at the wage of £7 a year—and no beer money.

The home must have been a blessing to the governors of local institutions, which were always overflowing with the debris of unhappy households. Although the girls had a hard life they did at least reap the benefit of decent occupations in an age when unemployment was the rule rather than the exception. The foundations of the laundry will probably remain in the paddock for all time, for they go deep into the local soil.

Dress

Introduction

England does not have a distinctive national dress that was worn with pride Sunday after Sunday. Mother's lace cap or shawl was not eyed with admiration or even envy: it was merely the badge of age, the relic of a bygone fashion. In the attics of large houses the clothes of earlier centuries have survived in trunks where they were packed away by the wearer, often for sentimental reasons. In the poorer families little survived the hard wear: a christening robe, or a child's first pair of shoes, but anything as once common-place as a pair of working breeches or trousers is a rare find. Shoes, on the other hand, were considered lucky and were some-times placed under the floor-boards of an altered or newly built house. Smocks, worn for some hundred years, survive; but it was the best, rather than the working smock, which was treasured.

Overshoes and pattens

When the living quarters over Burford's cycle and motor acces-sories shop were modernised by Brian Keylock, he found a wooden mousetrap and twenty wooden shoe-shapes between the two skins of a lath-and-plaster room-divider. The shape illustrated below is made of walnut and only 8¼in long. Miss Swann of Northampton's Central Museum, which specialises in footwear, identified it as part of a clog worn by a woman beneath her fine shoe in muddy weather. 'These wooden parts were usually covered with leather', she writes, 'and had a pair of straps over the instep, to tie the clog to the shoe. The straps were either of leather or, more usually, of the same fabric as the shoe. This style, with narrow squared-off toe and socket for a fairly low heel, dates from 1665–80. The blunt-pointed toe appeared in 1670, to become common by 1680, when it was combined with a higher heel. Finds of footwear concealed in the fabric of buildings are quite common in the southern parts of England and Wales; some strays have turned up even in North America. The shoes range in date from the early fifteenth century to 1934, and almost all were worn out, or deliberately slashed, when concealed. Nineteenth-century country pattens occur quite often and we have one clog part—like the Burford one—which was found under bedroom floorboard at a former inn at Blisworth, Northants. Presumably they were thought to bring good luck to an altered or newly built house.'

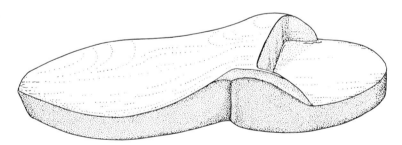

Above right is a pair of leather 'over-shoes' now in the village museum at Filkins, Oxfordshire. In style they exactly resemble a pair in the Gallery of English Costume, Manchester (also illustrated). As will be seen, they fitted over shoes, without necessarily giving them much protection. Miss Buck, Keeper of the Gallery,

prefers to call them clogs and sent this quotation from 'A Trip
Through the Town, 1735': 'Her neats-leather shoes are now trans-
formed into laced ones with high heels; her yarn stockings are turned
into fine worsted ones with silk clocks; and her high wooden
pattens are kicked away for leathern clogs . . . Plain Country Jane
is now turned into a fine London Madam.' This kind of clog seems
to have been bought with shoes, many having matching straps of
fabric or leather, and to have gone out of fashion about 1770.

The wooden mud-shoes illustrated on this page were sent by H. C. Hughes, of Burnham Overy Staithe, Norfolk. They measure 11in by 9½in and have crossed slats nailed to the undersides. The wearer apparently placed his shoes in the frames on top and held the strings to keep them in place. Doubtless they were used for walking across the mud-flats that abound along the Norfolk coast.

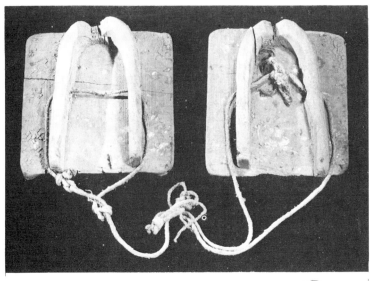

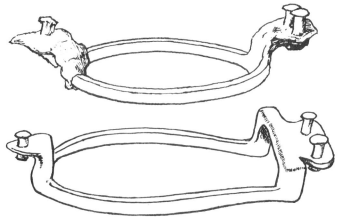

'Men are desired to scrape their shoes and the women to take off their pattens before entering this church.' N. Cufflin copied this notice displayed in the porch of the church at Stoke Albany in Northamptonshire. The wooden sole of the patten, a form of over-shoe, was riveted to the iron, which lifted the wearer clear of the mud in lanes and yards. R. T. Lattey of Neat Enstone, Oxfordshire, sent a drawing of a less common patten-iron reproduced left with another of the usual design. The special type may have been devised to give the wearer greater stability.

Pusher

The photograph (below left) of a wooden object with a spike $12\frac{1}{4}$in long came by way of the Castle Museum, York. The Curator writes that 'the curve (10in across) might fit the waist of a reasonably slim person'. For Alice Brebner it recalled a scene from the past: 'When I was a child, it may have been in the early nineties, I remember the then Lady Ashburton of Lochluichart being pushed slowly up a hill by her butler, who used a curved

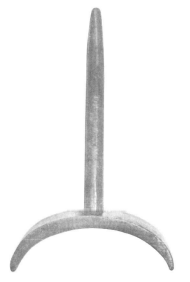

ebony band very like the one illustrated. She refused to allow her horses to tackle the steep road to the church. As she was immensely stout, it must have been quite an effort for the butler.' At Cliveden Antony Jarvis saw something similar but with a handle twice as long. Placed in the small of the back, it was used in the 19th century to give the wife of the then owner a helping push on the long climb from the temple by the Thames back to the house.

Bonnets

The photograph on the previous page was taken at a wedding in 1900. The women in their bonnets were about fifty years of age. G. W. Young of Colchester, an Edwardian gardener, tells us that girls and unmarried women preferred boaters. 'The reason for this, I think, was that bonnets worn by young women marked them as domestic servants. They were insisted upon by employers of two or more servants. As soon as a woman got married she started to wear a bonnet. I sometimes wondered if this was by choice, or at her husband's request to show that she was a married woman, and incidently his servant. Bonnets were mostly in dark colours, trimmed with ribbon and relieved by white artificial flowers. No flowers were allowed on those worn by servants.'

A gathering of smocks

We were often asked whether there was a connection between the embroidered design on a smock and the occupation of the wearer. There is no evidence of this on two recently presented to museums from the Burford area. One (opposite left), which survived in a village dressing-up box, is heavily embroidered with a flower motif. It belonged to a shepherd who was an old man at the turn of the century. The other (opposite right), of light fine linen, with round collar and simpler needlework, was the prized possession of a family of dairy farmers, the Clares of Clanfield. It is exquisitely darned and patched where the linen thread has perished. Not even the older members of the family recall any talk of when it was last worn, though they remember the care with which it was handled.

We put the point to Anne Buck, who has done considerable work on smocks at the Gallery of English Costume in Manchester; it was summarised in 'Folk Life', Volume I (1963). The smock frock, she tells us, developed its decorated form at the end of the 18th century, was at its best about the time of the Great Exhibition, then dwindled away in this century into the stockman's drill, then finally nylon, coat. It seems probable that there were many more drab, green and blue working smocks than are represented by survivals. Just as the stockman of today keeps his most dazzlingly white coat for the show ring, so it was the light-linen near-white smock which was most sparingly worn and carefully kept. (The only worn, torn smock we have seen is in heavy drill.) Among the survivors are wedding smocks, handed down from father to son, and Sunday smocks, perhaps lovingly worked for a child or sweetheart with emblems suited to his skill. But among the bought smocks it has not been possible to establish any connection between design and occupation. Many were made on cottage industry lines, one woman marking the patterns on all the work given out in a village; there are in existence metal blocks used for marking smocks mass-produced at Newark-upon-Trent. A widow earning her living by making corded bonnets put her own signature on the needlework, and the same may hold true for smocks. It appears that, at hiring fairs, some idea of a man's calling could be had from his smock or perhaps by the fact of his wearing one at all, though a tuft of wool, a piece of whipcord and so on were also worn to put the matter beyond doubt. Possibly the general cut was distinguishable. Did the shepherd, out in all weathers, wear a more ample garment, with large cape-like collar and four thicknesses of material across the shoulders? And did the cowman, working in a shed much of the time, sitting on a stool, wear a shorter one? A stone mason's smock we have seen has very little gathering; it would have been nothing but a nuisance holding the stone dust. But there are so few smocks with a fully recorded history that it is difficult to be sure of anything. We asked for sketches of any about which something was known: where and when it was bought, the occupation of the wearer and his place of work.

The request, Ann Cripps wrote, brought a welcome response. We were looking for regional period or occupational types, but our first impression was of bewildering variety. Did these frocks have nothing in common? We found in all of them only the square under-arm gusset, an absence of curves in the pattern pieces and a full width of material in the body. But Jane Woodland kindly invited me to see eight smocks which, ten or more years ago, were kept at the blacksmith's at Warbleton in Sussex and worn by the bearers at funerals. At Battle, in the same county, I was shown an 1838 garment from Crowhurst Park estate presented to the museum by Mrs Watson as a 'Sunday or funeral smock'; and Pelham Wait put on for me the frock he wears on Good Friday for the traditional game of marbles outside the abbey gateway. Rennie Bere had suggested a visit to Brighton Museum, where I saw a black smock made at Poynings in 1909 and presented in the same year by Ernest Robinson, member of a well-known farming family. We have also received from Irene Turner a sketch of a smock handed down by her grandfather, who was born in 1813 and farmed at Byworth, near Petworth.

On none of these Sussex garments was the embroidery on collar or cuff identical; but all had only a very small area of gathering on either side of the buttoned opening—on one it was no more than 1½in deep—and extra wide yokes to take the remaining width of material. The yoke, up to 11in wide, fell well over the point of the shoulder, so that the sleeve fullness sprang from the forearm, giving a characteristic outline sometimes just visible in local pastoral oils of Victorian days. None had embroidered boxes, and the needlework on yoke and cuff formed straight decorative lines. Only on the Crowhurst garment was there an inconspicuous heart to strengthen the bottom of the neck opening, and on Mr Wait's a reinforcing lover's bow. Here was a strong suggestion of a Sussex smock, simple and close in cut to a Regency shirt (below left).

The Sunday smocks were made from fine white linen, but at Battle I was told that some frocks were dipped to make them water-proof. Although it was impossible to be sure, the black unworn one at Brighton, which was not labelled 'Sunday', had the feel of a material that had been dressed. It was stiff, with a slightly shiny surface, and cotton and linen in their natural states are neither.

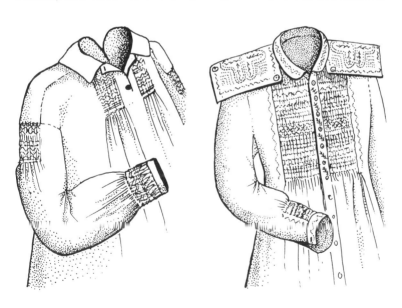

Would it have washed to the mole-grey of a smock from Warbleton? Was this a workaday version?

An impression that a regional style existed was current in the
early years of this century. Three sent and one detailed to us from
Shropshire had much in common (opposite right). N. Culliss sent
the working smock worn on Lyth Hill by her father David Barrett,
who died in 1935. It is made of stout twill and, with the $8\frac{1}{2}$in
collar, has four thicknesses over the shoulders; it must have been
a weather-defying garment.

From G. H. Wright we received photographs of the smock-
coat worn by William Head (1830–1911) in the Bridgnorth–
Shifnal area. It had a button-through opening down the front,
a tiny collar and large additional shoulder-pieces which projected
well over the points of the shoulders. Between each button there
is a diagonal sprig of embroidery, and centre back, at the nape of
the neck, two leaves and a flower; the overall needlework design
on the shoulder-piece is based on the heart. All these features are to
be found on the lad's smock, sent by Mrs D. M. Keeling Roberts
senior, which was the work of a smock-maker in the Ludlow or
Church Stretton district about 1900. Here the shoulder-piece has
decorative buttons. A book on 'English Smocks', written by Alice
Armes and owing much to the recording of Rennie Bere's mother
and great uncle J. E. Acland, also shows this overall heart pattern
on a buttoned collar-piece from Shropshire; and smocks in
Hereford Museum display strong similarities in cut and decoration.
So an outline quite different from that in Sussex emerges: a stout
drill garment having square emphasised shoulders, heavily
embroidered with heart motif.

A number of readers wrote to tell us of smocks they remembered
from childhood, with embroidery on box, collar and shoulder
yoke, in Surrey, Wiltshire, Devon, Buckinghamshire, Berkshire
and Oxfordshire. If the shoulder yoke had needlework, the collar
was probably short, so as not to hide the embroidery entirely; but
we have not enough information from any one county to draw
further geographical conclusions.

'I remember the carters coming in from Ashburnham with their

teams, the harness bells ringing, hooves clattering. The men wore
cords yoked up below the knee and smocks that just covered their
loins, down to here'—Mr Wait put his hand to the base of his
spine. Did all carters wear such short smocks, we wondered. It is
difficult to establish cut from a smock in the hand, for much
depends on the build and height of the wearer; so we turned to
contemporary sources. On a mug made in Newcastle about 1800
and titled 'Beating up for Recruits', the sergeant hands the long
knitted purse to a moon-faced fellow with a greedy glint in his eye.
The foolish one carries his carter's whip, a kerchief is knotted
round his neck, and his smock comes almost to his ankles. On a
1780 'God Speed the Plough' mug the man at the handles wears a
gown that comes almost to the back of the knees. It was always
popular to decorate architectural studies with amusing street
scenes—the dandy, the beggar, the smart chaise—and in nineteenth-century examples we have seen, the laden cart up from the
country is led by a man in a smock just above the knees.

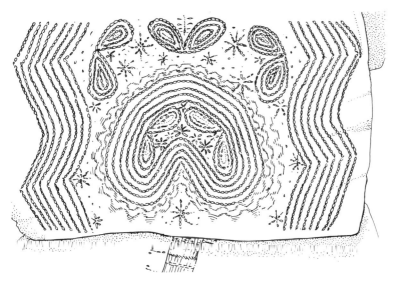

We have received details of five smocks worn by carters or
horsemen. Was the embroidery on them distinctive? Mrs F. E.

Welchman sketched the geometric pattern worked in white thread on the coarse blue linen of the smock of John King, ox carter, who worked all his life at Beckhampton in Wiltshire, until his death in 1917 at the age of eighty-two. The smock of a Berkshire horse coper, at least a hundred years old and lent by his grand-daughter Margaret Booker, has four main symbols embroidered in various combinations and sizes (below left). One, at first glance, might suggest a wheel; but in every use it is a tightly coiled spiral, not a circle, and this garment has the charm of a slightly home-made look. The Shropshire smock of William Head, waggoner, was probably made by his wife between 1835 and 1890. The units of the design are a heart, a flower and a pair of leaves (on the previous page). The fifth smock, a rough worn affair, has no embroidery other than the work on the tubing or gathering.

No characteristic cut or embroidery, which would have made a carter's smock recognisable at a glance to a hiring farmer, has yet

emerged. In *Under the Greenwood Tree*, which Thomas Hardy described as 'A Rural Painting of the Dutch School', he draws the scene in loving detail. The older carol singers and musicians 'wore thick coats, with stiff perpendicular collars, and coloured handkerchiefs wound round and round the neck till the end came to hand, over all which they just showed their ears and noses, like people looking over a wall. The remainder, stalwart ruddy men and boys, were dressed mainly in snow-white smock-frocks, embroidered upon the shoulders and breasts, in ornamental forms of hearts, diamonds and zigzags.' He makes no mention of an embroidered trade sign either here or in the hiring scene in *Far from the Madding Crowd*, where he writes: 'Carters and waggoners were distinguished by having a piece of whipcord twisted round their hats; thatchers wore a fragment of woven straw; shepherds held their sheep-crooks in their hands: and thus the situation required was known to the hirers at a glance'.

This does not mean, of course, that no wife or sweetheart was ever inspired by her man's calling or by some outside event; but these were perhaps the exceptions. Meanwhile E. Adelaide Ellery has guided us to a smock with a flower, shamrock and thistle, which may have been embroidered in a jubilee year (near left). L. John Mayes, Curator of High Wycombe Museum, where it is now preserved, writes: 'I have an extremely clear (but unauthenticated) memory of an old Hertfordshire shepherd telling me that his smocks were bought plain and decorated by his womenfolk. He had crossed crooks worked on the shoulders'. A great many smocks must have been made for unknown purchasers, with poverty as the spur. Some years ago an eighty-eight-year-old Berkshire labourer talked of his childhood to Judith Masefield: 'I did use to have to walk from Wheatley to the factory at Oxford with a bundle of those heavy smocks as our Mam worked in crinkle-crankle stitch. Blue, black and white milking smocks as I'd waxed the thread for and ruled the lines on in chalk. Mam only fetched in ninepence a smock.'

L. A. Reed has in her possession two family smocks. 'Tradition has it that the finer, in white linen, was worn by my great-great-

grandfather Jesse Colman at his wedding and on Sundays about
1830–40. He worked on the land, mainly with horses. The second
smock, tucked and not gathered, was made for my father by Miss
Hancock, now aged 90. My grandmother Mrs Drewitt made
milking smocks to just this pattern, she told me. The material
(holland) cost 4½d a yard, and she charged 9d for the making. Miss
Carr, who also lives here in the village of Edington near Westbury,
has the smock her carpenter-handyman father wore. It was made
by her grandmother, the cut being the same as for the others,
though it had the more usual gathering. Miss Carr recalls how
she and her brother used to set out with her grandmother's smocks
and a packet of bread and cheese to walk the eight miles to the
Melksham shop where they were sold. She was then eight years of
age and is now in her 70s. It seems likely that, at the end of the last
century, one pattern was in use in the village: the length of collar,
yoke, depth of embroidery and distinctive elongated triangular
gusset set in the side seam are common to all. Only the embroidery
varied with the maker's skill and the price the market would stand.'

In the 1930 edition of *The History of Speldhurst*, in an added
chapter, D. James describes an amusing incident during the
incumbency of the book's author, the Rev D. Mackinnon. When
the latter first arrived in the parish as curate in 1879 he found
that the men of the choir filed into church in their Sunday coats, and
the boys in smock frocks. So he supplied them all with surplices
at his own expense. All but one of the men went on strike and
boycotted the services; but the boys were delighted with what they
regarded as the latest London fashion in smocks.

Strong Drink and Good Company

Introduction

Changes in the village 'local' may be inevitable, but they result in a sad diminution of social life. The casual trade, with its bright chatter and smart cars, may have an air of vigorous life, but it bears no comparison with the strong companionships of a village community, the absorbing interest of 'darts night'. Games like quoits, skittles and shove-halfpenny need skill and are more satisfying than working a fruit machine.

   In the old days beer or cider—sometimes brewed on the premises —and companionship were not the only attractions: the pub might be the only place to get warm. A hundred years ago some cottagers could not afford to light a fire every day, and the cold of an un-heated stone building must have been almost unbearable to a man wearing his one pair of damp working trousers. For some the attractions of the local were more basic, and drunkenness was common.

Cider making

At one time cider mills and presses were commonly found in the

West Country, where farmers made cider for themselves and usually allowed their men to do likewise. The apples were pulped, and the juice was then squeezed out in a press. The two photographs below and right were taken at Rosenun Farm, St Keyne, Cornwall, by Rex Wailes. He wrote: 'This gear, used to drive the mill, is of the toothed roller variety; it is very primitive, being a face gear driving a lantern pinion on the shaft. The press is remarkable. The table and trough, into which the juice runs, were each cut from a solid piece of granite. The cake on the table consists of alternate layers of hay and crushed apples.' In some parts of the country—Herefordshire, for example—horsehair was used instead of hay, and in more recent times even coconut fibre. The purpose of the layers of fibrous material is to filter out the apple solids from the juice. In the picture of the press the wooden mill is partly obscured by the lefthand upright. The apples are fed from the loft floor into a hopper and fall on to two rotating spiked rollers, which crush them. A simpler form of mill can be seen on many farms, though not often in use. It consists of a circular stone trough, 8–10ft in diameter. The horse-gear drives a large stone wheel round and round to crush the apples in the trough. The photographs on page 186 were taken in 1953 in Herefordshire.

Jars and bottles

Dr Stuart Barber, of Buxton, inquired as to the most likely use for the three earthenware jars which he photographed outside a farm-house in Cardiganshire. They were found in a loft by the owner, whose widowed mother had lived on the farm all her married life and had not previously seen them; she was sure that they must be at least a hundred years old. Two are between 2ft and 3ft in height and hold about two gallons; the smallest one is some 18in high and holds half that quantity. The upper two-thirds of each is covered with a brown glaze, and all have bungholes at the base. Ffransis Payne, of the Welsh Folk Museum, thought they were used for the storage of water. Two readers suggested that they were most commonly used for home-made wine. M. J. Gladwell, of Sheffield, reported the finding of a similar jar in his great-grandmother's wine cellar. From Cardiganshire, where the published photograph was taken, Mrs E. M. Davies, of Llanon, wrote that she had such a jar as a gift from her uncle. A neighbour had told her that they were used mostly for making wine at home, and that her mother used them for ginger beer in the winter time, when the cow was dry and there was no bread-and-milk for her large family's supper. She made the ginger beer in a round nine-gallon iron pan, which was hung over the fire. She cut off the top of a sheaf of barley or, when this was not available, of mixed corn, and simmered it for hours with little bits of herbs, of which she always had a stock ready dried under the roof. She added hops and brown sugar, and finally ginger and yeast when it was cool. The brew was then put into the jar, which had a small wood tap at the bottom in the bung-hole. For supper, toast was cut up and put into basins, the warmed ginger beer being poured over it. The children evidently went to bed thoroughly warmed and slept like logs after it. Occasionally, Mrs Davies was informed, such jars would be used to make butter when the cow was nearly dry and there was very little cream available. A wooden lid was fitted over the top, and a wooden handle with a flat piece of wood on the bottom was worked up and down through a hole in the lid.

The earthenware bottle, seen in close-up below, was used for taking cider into the fields. It came from J. J. Rossiter of Corsham, Wiltshire. It is a well-designed specimen, $8\frac{1}{2}$in high.

Well and wheel

The photograph below was taken in 1955 by W. R. Bawden. F. W. Hoar, landlord of the Fox and Hounds Inn at Beauworth, Hampshire, is seen treading this wheel as he had done for 41 years; formerly it was trodden by a donkey. He had to walk 794yd to draw 18 gallons of water from a depth of 300ft. The well dates from the 12th century. In the days when pubs brewed their own ale a good supply of water was of prime importance.

Wooden hammer

The 7½in hammer-like tool above was found by Anthony Blatch-ford at Northcott Mouth, Devon. It has Belgian studs in the head. It suggested a meat tenderiser to some readers, but the light weight of the head ruled out this possibility. George Swinford of Filkins in west Oxfordshire tells us that his family used such a hammer in making rhubarb wine. Acid liquids must be kept away from iron, so the rhubarb fibre was broken down with a wooden hammer on a little wooden stool—the anvil—stood in a tub. Juice and bruised stalks went into the tub; boiling water was poured on, and the whole was left to stand for two or three days before straining.

Bottle-boot

The object (opposite above), open at both ends, was brought by Jack Cross, a Saffron Walden farmer, who found it lying in a barn. It is made of thick leather and at the wider end has a metal frame with a projection on either side. It is 8½in long and has a diameter of 4in at the larger end of 2in at the other. We thought of the object as funnel-shaped and published a photograph of it upside down, but Miss A. Nicholl of Buntingford had no difficulty in identifying it. She had in her possession the bottle-boot used by her grandfather in Essex. A. L. Morell, of Nottingham, knew a farmer who owned one. He explained its use thus: 'The boot was placed over the bottle; a corkscrew was then inserted, the feet were placed on the projections and a hand on the neck, and out came the cork'. C. E. Freeman has a specimen in Luton Museum, and the donor, T. W. Bagshawe of Angmering-on-Sea, wrote that it was used to hold the bottle while the cork was being inserted, for protection in the event of a burst.

Bottle opener

The blade of the tool below, sent by S. F. Forrester of Knutsford, Cheshire, is $1\frac{1}{2}$in long, and the spike protrudes the same distance beyond it. Its appearance suggested some association with net making or rope work, but D. Carter, of Christchurch in Hamp-

shire, told us that fifty-five years ago they used a similar tool to tend the bedroom candles. If one burned unevenly, the top would be flattened with the blade, which was also used as a snuffer. The point was used to prize the candle-end from the stick. J. B. Nevitt, of Pulborough in Sussex, remembered his father using an identical tool sixty years ago, the blade for removing the wax seals from wine bottles and the spike for digging out recalcitrant pieces of cork. R. A. Salaman told us that such 'bottle-openers' were commonly used to uncork champagne: the blade for cutting off the wax and the spike for removing the wire and easing out the cork. He sent a drawing of a very similar tool from a catalogue dated 1892, the only difference being that the blade was there serrated. Mr Carter's tool may have been designed as a bottle-opener and later used for candles.

Toddy stirrers

The group below shows three toddy stirrers received from Miss Elizabeth Atkinson of Berrier, Cumberland. The metal stirrer on the right was one of a set belonging to a Westmorland innkeeper. The other two are made of glass: the rougher one in the centre was bought at a cottage sale, and the one on the left with a twist in it and of a generally more decorative design was among the effects of the owner of the 'big house' in the same village.

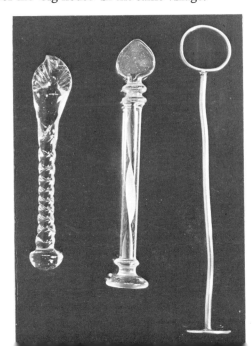

Nutmeg grater

The nutmeg grater below, shown open and closed, is about $3\frac{1}{2}$in high, and belongs to Mrs W. Mayne of Worthing, who wrote, 'When my father (born in 1840) was a young married man, an elderly aunt lived with him. She liked to have with her supper a glass of ale which she would flavour with nutmeg from this little grater kept in her reticule.'

Puzzle

The interlocking rings, found in a wheelwright's shop near Swindon, have been identified by G. L. E. Turner, of the Museum of the

History of Science, as the ancient Chinese rings puzzle. In the picture above the puzzle has been solved, the rings having been removed from a narrow iron loop. Some hundreds of moves, casting the rings off and on the loop, are required to unravel it. In *Amusements in Mathematics* (1958) H. E. Dudeney wrote that 'it is said still to be found in obscure English villages, sometimes deposited in a strange place, such as a church belfry . . . and to be used by the Norwegians to this day as a lock for boxes and bags'. These 'tiring irons' had a vogue at the end of the last century around Swindon and were probably made by men working in the railway shops there.

Pins and skittles

A correspondence about skittles and wooden cheeses took us down interesting alleys. In the 14th century 'kayles' was a popular pastime. Pins were set up in a row, and the players tried to knock them down with a thrown stick. The French called it *jeux de quilles à baston*; so quilles, kayles, kittlepins, skittles. In Strutt's 'Sports and Pastimes' cloish or closh is described as being similar to kayles, but 'the pins were thrown at with a bowl instead of a truncheon'. Variants of the game developed in which a ball or bowl was rolled along the ground—ninepins is one—but in skittles the original throwing action was retained. Some 60 years ago many country pubs had their skittle alleys, and we have heard of some in Dorset, Somerset and Gloucestershire where the missile was a solid

dumpy cheese shape, about 10in across and 4in thick. One correspondent, E. J. Smith, recalls seeing the game being played at a
pub by the quay at Poole in the 1930s. The cheese was delivered
two-handed, and the whole-hearted player fell full length on the
ground as he let go. The atmosphere was hilarious, and there was
much drinking and betting. One of the few remaining full-length
alleys is run, more sedately, by the Westminster Bank Skittles
Club at Norbury. The Secretary tells us: 'The aim, as in golf, is a
low score—one point for each throw. You hope for a "floorer",
when all nine pins fall to one throw. The pins, made ideally of
hornbeam, weigh some 9lb each and are arranged in a diamond-
shaped frame with a bevelled edge an inch or two above floor
level. The 12lb cheese is thrown 21ft and must strike the pins full
toss. It is discus-shaped, 12in across and $3\frac{1}{2}$in through the hub,
and made of *lignum vitae*, the only wood that will stand up to the
rough usage.' Other correspondents have seen the game played at
Mortlake and Putney, where Norman Tyrer went with a rowing
friend in the 1920s. 'The player held the cheese between his
curved hand and forearm', he writes. 'The pitching required
considerable effort and a good sense of balance. We regarded
skittles as the real thing and rather looked down on the game of
ninepins.' J. W. Stevens and R. Sharp both describe a scaled-
down table game of skittles. It is said to have originated with the
great-grandfather of W. T. Black, who makes tables, like the
one sketched below to this day at Northampton. Village teams

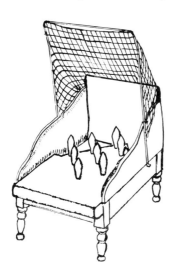

compete against each other in the county, and tables have been encountered in Gloucestershire and Devon. The cheese, a disc about 3½in across and 1½in thick, is thrown 10ft underarm with a skimming motion; it may hit the pins directly or after bouncing on the table. 'A low throw which hits the front of the table may go anywhere and does.' Each player throws three cheeses, and the team that knocks down most pins wins the 'horse' or 'chalk'; the first team to win five 'horses' takes the game.

Nine men's morris

A. C. Hilton wrote about one of the oldest games to be found in England, nine men's morris. 'It persisted in my own family quite strongly until a few years ago. Even now, at Christmas reunions, we sometimes test each other out. I have heard it said that the Vikings introduced it into this country. Whether this be so or not, it is certain that the old wherrymen of East Anglia played it regularly. It is mentioned by Titania in "A Midsummer Night's Dream" in the lines:

> "The nine men's morris is fill'd up with mud;
> And the quaint mazes in the wanton green,
> For lack of tread, are undistinguishable."

The glossaries of some editions confuse it with the morris dance, while others say that it is a game. I believe that it was played on village greens, on wooden boards or stone slabs, and possibly, too, on patches of bare earth. There is evidence for this at certain places in the form of incised stones, such as that in Wales above Llanfairfechan, near the meeting of four tracks, of which one is a Roman road. Records show that the shepherds of Salisbury Plain knew how to play it.

The game can be played anywhere, at any time, without highly priced or skillfully carved pieces. Buttons of different shapes or colours, pieces of stick of varied lengths, stones, hips and haws— almost any improvisation will serve. The board may be rectangular or square, with lines as shown in the diagram above right. The circles indicate the points which the men may occupy.

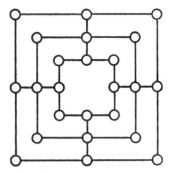

Two players, each with nine men, take turns to place them, one at a time, at the intersections of the lines or at the corners. When the board is set, six points remain unoccupied. The aim of each player is to secure a row of three men on any line on the board, and at the same time to prevent his opponent from doing so, by moving one of his own men into a line which looks like being "made". A man may be moved along a line to any vacant point immediately next to the one it occupies; no jumping is allowed. The skill lies in manoeuvring the men so that a player can open up a row and close another by moving a single piece. As soon as a row of three has been made, the successful player can take any one of his opponent's men which does not form part of a closed row of three at the time.

This goes on until one player has only three men left. He is then allowed to jump to any position on the board, but he usually does this without much success. The game is up when a player has only two men left and so cannot form a row or line. When both have three men left and so can prevent a line being formed anywhere, it is a draw. In the matter of skill, nine men's morris is on a level with draughts'.

Quoits

The photograph overleaf, taken in the early years of the century, shows a group of Suffolk quoit players. Alan Richardson gave us this background information. 'Quoits is said to have originated in the border country between England and Scotland. It was known, and

prohibited, in the reign of Edward III. In 1545 Roger Ascham
wrote, "Quoiting be too vile for scholars". The game was familiar
to me during my boyhood in North Yorkshire, and I believe the
depression after the 1914–18 war, when men needed inexpensive
pastimes, caused a revival of interest in the industrial towns. It
seems no longer to be played there, but in the Esk valley near
Whitby at least a dozen villages have teams, and many more have
pitches [1956].

A quoits club is neither highly organised nor exclusive. For a
pitch a piece of flat ground 15yd long is needed, and it is not
difficult to find in a moorland area where much of the land is un-
fenced. At Beck Hole, Goathland, the pitches are on the green; at

Danby they are beside a wide grass lane behind the village street, and at Goldsborough a suitable place has been found at the roadside. The small amount of necessary equipment can be kept in a handy barn or garden shed, a public house or the dwelling of a member who lives near by. Pitches usually go in pairs, side by side. A pitch has a "box" about 4ft square, excavated at each end and raised again to ground level with clay which is kept in a moist putty-like condition. The centres of the boxes are 11yd apart; their edges are preserved with stout wooden planks, and the first two yards of the pitch at each end, from which the players throw their quoits, are also floored with wood.

Those are the essentials. An enthusiastic club may also provide rough seats for players waiting their turn, a scoreboard between the pitches, and a high rail for spectators to lean on. The area round the pitches and between the boxes receives hard wear. It may be turf, kept short by sheep, or it may have been concreted or top-dressed with stones or gravel. When the pitches are not in use the clay in the boxes is covered with wet sacks, and a cover is placed over each box so that the clay does not dry out.

Matches, made up of games between two players, usually take place in the evenings. Each player in a match has two heavy iron quoits, weighing between 5lb and 7lb and shaped like a deep saucer with the flat centre cut out. From the boards in front of a box the players throw their quoits alternately at the hob in the box at the other end. The hob is a long iron rod driven straight down into the clay in the centre of the box until it protrudes no more than 4in. When the four quoits have been thrown a player counts a point for each of his quoits which is nearer to the hob than either of his opponent's; a "ringer", a quoit completely encircling the hob, counts two. The player winning an end has first throw in the next, and the first to score twenty-one points wins the game.

Throwing ringers is not difficult for a good player, but only against a weak player is it usual to attempt a ringer with the first shot; a good opponent might cancel it with a second ringer on top of the first, and there is rarely room for a third. The quoits are usually thrown to stick on edge in the soft clay, and the shape of the quoit is used to advantage to present an opponent with an obstacle from which his quoit will glance into an unfavourable position. The shots which can be played, either to foil an opponent or to squeeze into a narrow gap, are named according to whether the quoit lodges to left or right of the hob, and whether it rests with the rim facing up or down. The main shots are, on the left of the hob, the "side quoit" if the rim is towards the ground, and the "pot" if it is the other way up. Corresponding shots on the right of the hob are the "cue" (Q?) and the "Frenchman". A quoit on edge directly in front of the hob is a "gater". This is a popular first shot, as it defends the hob and may be knocked down into a ringer at the next throw.

The games often go on into the dusk and even the darkness. At such times the hob will be chalked, and a team-mate of the thrower may hold up a piece of chalk or a match-box to indicate the critical tenth of an inch of clay and show him how best to place his quoit.'

Acknowledgements

Drawings by Rosemary Wise with the exceptions of John Anstee 17, 39, 48, 78 (top), 96, 99; V. R. Sharp 38, 82, 131; Brian Walker 67, 68, 113; A. C. Scarrott 86; James Arnold 89, 91; T. D. Parks 130; R. T. Lattey 172; R. Sharp 195.

Photographs by the Museum of English Rural Life with the exceptions of Douglas Hughes 14, 15, 16; Kathleen Holter 23; M. Wight 25; John Gilbert 30, 85 (below); Christine Wright 34; E. Murray Speakman 35 (left), 57 (right); John Peterson 37; British Museum 42; H. E. Tyndale 44; W. T. Jones 47, 81 (left), 97, 132, 137, 138, 171 (top); Eric Cheek 53, 55 (left); C. Henry Warren 57 (left); Gordon Clemetson 59 (top); Walmsley & Webb 60 (top), 73 (top); Levi Fox 61; W. R. Mitchell 62; Douglas Went 63; D. Jeffery 81 (right); J. E. Skyrme 83; G. D. Bates 87; E. O. Hoppe 95 (below); John Cripps 98; W. A. Prevost 103; *Braintree & Witham Times* 105; S. E. Godwin 107; J. O. Evans 109; B. R. Townend 110, E. H. D. Williams 111; L. Sanders 117–22; Jack Hill 122 (below); Douglas West 125; K. H. Bentley 133; M. I. Crowther 134; Allan Jobson 135, 198; Ulster Folk Museum 136; Cambridge & County Folk Museum 144; G. C. Farmer 160 (left); W. A. Cocks 163 (right); K. T. Stevenson 164; R. Cripps 165; Gallery of English

Costume, Manchester 171 (below); Castle Museum, York 173;
G. W. Young 173 (right); Rex Wailes 184, 185; M. F. Collins 186;
H. Barber 187; W. R. Bawden 189.

Items illustrated on the dustjacket

Bee vaccinator: It has apparently been believed for a very long time
that bee stings assist in the care of rheumatism; indeed, even now
one leading London hospital buys bee venom for experimental use.
The bee vaccinator is made of tinned metal with a small glass tube
mounted above. The base section has a slide which opens into this
tube. The bee is encouraged to go in through the hole and the slide
is then put back in position. In use, the vaccinator is put onto the
section of the skin where the sting is required; the little plunger is
pressed down gently, the bee is forced into contact with the skin and
hopefully the patient gets stung!

Blow-lamp: This early blow-lamp has a copper body, steel tubular
nozzle and iron folding handles. Because of the type of fuel used,
it is probably late Victorian or early Edwardian.

Candle-lamp: This late eighteenth or early nineteenth century
lamp is of French or Italian origin. It is made of tin-glazed earthenware and perforated vents are visible in the floral decoration at the
top.

Sausage machine: Made of English maple and lined with pewter,
this early mincing machine has two rows of eight teeth arranged in
a spiral, which push the meat against sharp steel knives and, by the
spiral action, eject the meat through the nozzle.

Index

Mill, bill, 58
cider, 183
hand, 58
steam, 59
stone, 59
Mould, curd, 145
gingerbread, 139
Mullers, paint, 103

Nail-drawer, 76
Needle, thatching, 96
Nine men's morris, 196
Nutcracker, 141
Nutmeg grater, 193

Oven, communal, 129
Oxen, 22

Pack saddle, 20
Paint muller, 103
Pap boat, 137
Patten, 169
Peg, slater's 97
Pegotty, 150
Pipe borers, 104
Plough, cleaner, 39
foot-, 34
share, 39
Potato hoe, 37
Press, cider, 183
hay, 63
Pusher, 173

Quorn, 131
Quoits, 197

Rings puzzle, 193

Salt box, 133

Saw, veterinary, 30
Scarifier, 28
Scoop, apple, 138
marrow, 139
Seedlip, 40
Shoe, over-, 169
mud, 172
Skewer holder, 132
Skittles, 194
Slate-iron, 99
Slat-pick, 98
Smock, 174
Splitter, slater's, 97
straw, 147
Spoke dog, 84
Staymaker, 152
Steelyard, 120
Strake, 81
Straw rope, 64
splitter, 147
Sugar cutters, 139

Tarigrep, 37
Teeth extractor, 29
Thatching needle, 96
Threshing, Burrell, 66
frame, 56
hand machine, 56
team, 52
Thwacker, 100
Toddy stirrers, 192
Tongs, blacksmith's, 80
weeding, 46
Trace link, 35
Traveller, 80
Turnip slicer, 61
Tuttle box, 18
Tyring wheels, 77